Empowering the Mentor

of the

Beginning Mathematics Teacher

Empowering the Mentor

of the

Mathematics Teacher

A series edited by Gwen Zimmermann

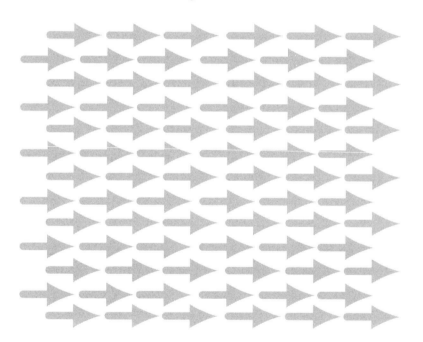

Empowering the Mentor

of the
Beginning Mathematics Teacher

Edited by

Gwen Zimmermann
Adlai E. Stevenson High School
Lincolnshire, Illinois

Patricia Guinee
Peoria Public Schools
Peoria, Illinois

Linda M. Fulmore
Mathematics and Equity Education Consultant
Cave Creek, Arizona

Elizabeth Murray
Navajo Elementary School
Albuquerque, New Mexico

NATIONAL COUNCIL OF
TEACHERS OF MATHEMATICS

Copyright © 2009 by
THE NATIONAL COUNCIL OF TEACHERS OF MATHEMATICS, INC.
1906 Association Drive, Reston, VA 20191-1502
(703) 620-9840; (800) 235-7566; www.nctm.org

Library of Congress Cataloging-in-Publication Data

Empowering the mentor of the beginning mathematics teacher / edited by Gwen Zimmermann,
Patricia Guinee, Linda M. Fulmore, and Elizabeth Murray.

p. cm. — (Empowering the mentor of the beginning mathematics teacher)

ISBN 978-0-87353-620-2

1. Mathematics teachers—Training of. 2. Mentoring in science. I. Zimmermann, Gwen. II.
Patricia Guinee. III. Linda M. Fulmore. IV. Elizabeth Murray.

QA10.5.E468 2009

510.71—dc22

2009003575

The National Council of Teachers of Mathematics is a public voice of mathematics education,
providing vision, leadership, and professional development to support teachers in ensuring
equitable mathematics learning of the highest quality for all students.

Printed in the United States of America

Contents

Contents

Preface

[S]eldom, if ever, do we ask the "who" question—who is the self that teaches? How does the quality of my selfhood form—or deform—the way I relate to my students, my subject, my colleague, my world? How can educational institutions sustain and deepen the selfhood from which good teaching comes?

—Parker J. Palmer
The Courage to Teach

All too often we hear anecdotes of teachers' leaving the field because they are overwhelmed by the demands of teaching. Perhaps a teacher education program has prepared a novice teacher with the necessary mathematics content knowledge, a foundation in pedagogy, some classroom discipline techniques, and hands-on experience in the classroom. Maybe a more experienced teacher is struggling to keep abreast of the constant barrage of changes in the field or within his or her own building. Even within the supportive structure of a university teacher preparation program, in-service teachers may feel weighed down by all the demands placed on them. Regardless whether one is a beginning, experienced, or preservice teacher, one can become overwhelmed by all that is required to merely survive let alone flourish as a mathematics teacher.

In the quote above, Parker Palmer challenges us to ask ourselves how we might help colleagues on their continuous journey to better their teaching. Do we leave our colleagues to flounder as they navigate all the complexities of what it means to teach mathematics? Mentoring is the answer to Palmer's question of how we might "sustain and deepen the selfhood from which good teaching comes." Mentoring can provide the support and encouragement not only to survive the demands and challenges of teaching but also to thrive and develop as professionals who are dedicated to the teaching of mathematics.

In 2004, NCTM published a series of publications titled *Empowering the Beginning Teacher of Mathematics.* Realizing that a gap existed in providing similar support specifically for mentors of mathematics teachers, NCTM's Educational Materials Committee issued a call for manuscripts that would provide the basis of practical "how to" advice for individuals who participate in formal or informal mentor training or serve in the capacity of instructional coach, peer coach, lead teacher, collaborative peer, department chair, administrator, critical friend, team leader, university supervisor, or department or grade-level colleague.

The original intent of the call was to create grade-level publications mirroring the framework of the beginning teacher books. However, when the editorial panel met to review the numerous submissions, the advice for mentors and mentoring programs was not so much differentiated by grade level but rather, was distinguished by the level of teaching experience of the teacher being mentored. The result is separate publications on the mentoring of beginning mathematics teachers, experienced mathematics teachers, and preservice mathematics teachers. Although some mentoring advice is specific to each group, other mentoring advice transcends any amount of teaching experience.

The intent of the editorial panel was to create publications on the mentoring of mathematics teachers that would be informative and practical resources that are easy to reference to address the reader's specific needs. We hope we have provided some useful ideas as well as challenged the reader to think differently about what it may mean to be a mentor.

REFERENCE

Palmer, Parker J. *The Courage to Teach: Exploring the Inner Landscape of a Teacher's Life.* San Francisco, Calif.: Jossey-Bass, 1998.

Mentoring New Teachers
A Position of the National Council of Teachers of Mathematics

States, provinces, school districts, and colleges and universities share responsibility for the continuing professional support of beginning teachers by providing them with a structured program of induction and mentoring. These programs should include opportunities for further development of mathematics content, pedagogy, and classroom management strategies.

The retention of new teachers continues to be a concern in both the United States and Canada. Statistics show that nearly half of the new teachers in the United States leave the profession in their first five years of teaching, and Canadian and U.S. attrition rates are both around 30 percent for teachers in their first three years. These high rates of attrition contribute to the overall shortage of high-quality mathematics teachers, particularly at the middle school and high school levels. This attrition is especially alarming in the United States, where it is predicted that more than 2 million new teachers will be needed in the coming decade.

In far too many schools, new mathematics teachers receive challenging teaching assignments for which they are unprepared. These teachers, some of whom do not have strong backgrounds in mathematics content, are often isolated from professional involvement with colleagues. Frequently, they receive little content-specific professional development to support them in meeting the challenges that they face. As a result, their students may not be afforded the learning opportunities and quality instruction that the Council advocates as essential preparation for high-functioning adults in the workplace and everyday life.

Recommendations

States, provinces, school districts, and colleges and universities should provide professional development for new teachers by creating partnerships between experienced and novice teachers. These partnerships should ensure a strong focus on mathematics content knowledge, pedagogical knowledge, and knowledge of Principles and Standards for School Mathematics (NCTM 2000) and its application to the classroom. Education agencies should establish mentoring programs for new teachers and provide funding for the programs and the training of mentors. In making teaching assignments, district and school-based administrators should consider the additional demands on beginning teachers and their mentors alike. Teachers who have been identified as mentors should receive significant and consistent training, as well as appropriate remuneration or release time for their services. Finally, beginning teachers need and deserve a strong, structured program of induction, which includes mentoring, to ensure their success and increase the likelihood that they will stay in teaching, growing steadily in professional expertise and finding lifelong satisfaction in a career of continued service to mathematics education.

(September 2007)

Introduction
Patrica Guinee

ROBERT Fulghum's book *All I Really Need to Know I Learned in Kindergarten* (1988) poses that the essentials for a well-balanced life are learned in the kindergarten classroom. Unfortunately, the same does not hold true when it comes to teaching mathematics. In fact, although teacher education programs can lay a foundation on which to build the knowledge and skills of teaching, they cannot provide the opportunities to learn everything one needs to know about teaching. Much of the development of the art and science of teaching is learned "on the job."

For years this essential component of learning has been left up to the individual classroom teacher, thus ensuring that the learning taking place was haphazard. Some individuals were lucky and were able to develop into accomplished teachers. Others were not so lucky, and their tenure in the ranks of teachers was brief. Recently, the educational community has recognized the need to mentor beginning teachers. As a result, much has been written about mentoring and many teacher induction programs are being established. Educational journals and school districts across the nation have begun to address the needs of beginning teachers; however, the need continues for specific support of the beginning teacher of mathematics.

Beginning teachers are making the transition from student to teacher. As elementary and high school students, many beginning teachers may have been taught mathematics as a set of procedures with a focus on arithmetic and symbolic manipulation. What can be done to help these beginning teachers focus on the development of mathematical concepts and connections for their students? How do we help beginning teachers create standards-based mathematics classrooms for their students? How can we support beginning teachers as they address the complexities of balancing the need to teach for depth of mathematical understanding with the pressure of preparing students for high-stakes testing?

Mentors of beginning mathematics teachers can address these specific needs. This book provides a platform for mentors and mentees to share their experience and advice for those who plan to mentor beginning mathematics teachers. Section 1, "Why Mentoring Is Important," lays the foundation for the need to mentor beginning mathematics teachers. Section 2, "Who a Mentor Is," and Section 3, "What a Mentor Does," discuss characteristics of mentors and suggest a wide variety of roles that mentors of beginning mathematics teachers might fill. Section 4, "Tools for Mentors," offers a number of helpful strategies for structuring conversations, conducting observations, and planning your mentoring experience. Section 5, "Collaboration as Mentoring," highlights the collaborative nature of mentoring. Section 6, "Ideas for Mentoring Programs," discusses existing programs and offers various models to consider in planning a mentoring program. Finally, Section 7, "Lessons Learned," shares mentors' reflections and invaluable insights gained in their mentoring experiences.

This book is a rich resource in which you will discover many applicable ideas. Whether you have prior mentoring experience or are beginning to consider being a mentor, you will find the information valuable. I invite you to take advantage of the thoughts and advice compiled here as you develop and refine your skill in mentoring the beginning mathematics teacher.

REFERENCE

Fulghum, Robert. *All I Really Need to Know I Learned in Kindergarten.* New York: Villiard Books, 1988.

Section 1:
Why Mentoring Is Important

REMEMBER your first year as a teacher? You had high hopes and expectations, both for yourself and your students. You were eager to be in charge of your own classroom. Just like you, today's beginning teachers seek the challenges of designing their own lessons, helping students learn mathematics, and growing as professionals. They are passionate about teaching and ready to begin what they view as a lifetime commitment.

Given the excitement and energy that beginning teachers bring to their first teaching positions, why is having a mentor important for newcomers to the profession? These teachers are fresh out of college and eager to bring new ideas into their classrooms. Yet often, the enthusiasm and passion evident at the beginning of the year give way to frustration and disillusionment by November. Mentoring, like teaching, is much more complex than merely helping teachers get past the "November dip." The articles in this section highlight the multifaceted nature of mentoring and the benefits mentoring has for beginning teachers.

66 Those having torches will pass them on to others. 99

—Plato

Mentoring Mathematics Teachers in the Twenty-first Century

With changing demographics in the United States, the need for effective mentoring of teachers, particularly mathematics teachers, has become increasingly important. Recent demographic data indicate that approximately 48.3 million children are enrolled in public schools in the United States in prekindergarten through twelfth grade. Of these students, 43 percent are members of a racial or ethnic minority group, 19 percent speak a language other than English at home, and 14 percent receive special education services (Rooney et al. 2006). Further, the U.S. Census Bureau (2006) reveals that about 18 percent of children under eighteen years old are living in poverty.

In the current era of high-stakes testing and accountability, a mentor must possess the knowledge, skills, and disposition necessary to provide effective mathematics instruction to all students, along with assistance to new teachers. Given the diverse, changing student demographics in the United States, the mentor and mentee's relationship must be grounded in the notion that "[a]ll students, regardless of their personal characteristics, backgrounds, or physical challenges, must have opportunities to study—and support to learn—mathematics" (NCTM 2000, p. 12). Further, an effective mentor is one who is knowledgeable about research in mathematics teaching and learning and best practices in the field.

The standards set forth by the Interstate New Teacher Assessment and Support Consortium (INTASC) provide a solid framework for mentors as they support, through deliberate scholarly planning and decision making, the development of beginning mathematics teachers. Table 1.1 highlights how the INTASC standards are articulated with respect to mathematics teaching and learning.

For beginning mathematics teachers, the mentor is a powerful and essential supporter and sponsor. The mentor functions as a bridge between teacher preparation, whether traditional or alternative, and new membership in the mathematics teaching community. Whether mentoring a first-year teacher or someone with fifteen years of experience, the goal is the same: to provide "more and better mathematics for all students."

— *Shonda Lemons-Smith*

REFERENCES

National Council of Teachers of Mathematics (NCTM). *Principles and Standards for School Mathematics.* Reston, Va.: NCTM, 2000.

Rooney, Patrick, William Hussar, Michael Planty, Susan Choy, Gillian Hamden-Thompson, Stephan Provasnik, and Mary Ann Fox. *The Condition of Education 2006.* National Center for Education Statistics Publication 2006071. Washington, D.C.: National Center for Education Statistics, 2006.

U. S. Census Bureau. *Current Population Survey.* 2006 Annual Social and Economic Supplement. Washington, D.C.: U.S. Census Bureau, 2006.

Mentoring Teachers in Standards-Based Mathematics Education: A Visual Framework

Most of us who have mentored teachers in mathematics education are aware of an inherent and perplexing challenge: "Teachers, equipped with vivid images to guide their actions, are inclined to teach just as they were taught" (Ball, as cited in Wolodko, Willson, and Johnson [2003, p. 243]). In mentoring preservice and in-service teachers toward Standards-based practices, I have found that much of the work involves helping new teachers develop personal images, metaphors, and references that may be different from those to which they are accustomed.

Although I share my visual framework for Standards-based mathematics learning and pedagogy as a reference (see fig. 1.1), I urge teachers with whom I work to develop their own representations. The images they create seem to capture their personal views of mathematics education while revealing some of their concerns, as well.

Typically, a useful framework includes only enough detail to remind the teacher of major organizing ideas, concepts, and relationships; too much detail may detract from the purpose of the framework. The relationships among the various parts of the framework can be made explicit through placement, arrows, colors, font size and style, and so on. The real value of such an illustration, however, is not what is communicated to others but the ideas, concepts, and relationships the image evokes for the person who constructed it. The framework can be altered or completely reorganized as a result of additional learning or new insights.

Table 1.1

INTASC (Interstate New Teacher Assessment and Support Consortium) Standards Applied to Mathematics Teaching and Learning

Elements of Practice in INTASC Standards	Application to K–12 Mathematics Teaching and Learning
Standard 1: Knowledge of subject matter	Mathematics teachers should possess strong content knowledge in number and operations, algebra, geometry, measurement, and data analysis and probability.
Standard 2: Knowledge of human development and learning	• Mathematics teachers should understand developmental factors that affect learning and ensure that mathematics instruction takes into account multiple intelligences.
Standard 3: Adaptation of instruction to individual needs	Mathematics instruction should— • be differentiated to meet the needs of students at various cognitive levels, including those who are struggling and those identified as gifted or talented; • reflect principles of culturally relevant pedagogy or culturally responsive teaching; and • take into account students for whom English is a second language and students who have special needs.
Standard 4: Multiple instructional strategies	• Mathematics instruction should effectively use manipulatives, technology, and other tools for learning, along with a variety of modalities, such as student-centered learning, cooperative learning, and small-group and whole-group activities.
Standard 5: Classroom motivation and management	• Teachers should foster a mathematics classroom culture in which all student voices are equally valued and encouraged to participate in the learning process.
Standard 6: Communication skills	• Mathematics instruction should consist of high-level problem posing and questioning, as well as teacher-student and student-student mathematical discourse.
Standard 7: Instructional planning	Mathematics instruction should— • reflect high expectations and consist of high-level mathematical tasks; • build on students' informal mathematical experiences, prior knowledge, strengths, and interests; and • be aligned with local, state, and national mathematics standards.
Standard 8: Assessment of student learning	Mathematics instruction should afford students flexibility and multiple opportunities to demonstrate their mathematical understanding.
Standard 9: Professional commitment and responsibility	Mathematics teachers should strive to be reflective practitioners who critically evaluate their roles in promoting students' mathematics achievement.
Standard 10: Partnerships	Mathematics teachers should participate in professional learning communities, such as school-based, local, state, or national mathematics organizations.

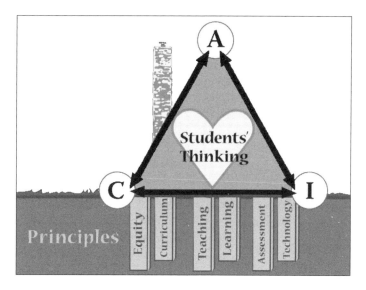

Fig. 1.1. Visual framework: My house of mathematics

For my visual framework, which continues to evolve, I have borrowed heavily from both my preparation and experiences as a teacher in various contexts, including the elementary school classroom situations in which I have served as a supervisor, professional development workshops, and university teacher education courses. The *Standards* documents of the National Council of Teachers of Mathematics (NCTM 1989, 1991, 1995, 2000), the writing and research of colleagues in the profession, and my experiences in supporting preservice and in-service teacher development have heavily influenced my model. The following paragraphs describe my framework, which I think of as a "house of mathematics."

The Foundation

Standards-based mathematics classrooms may look different from one another on the surface, but the knowledgeable visitor will see evidence of decision making guided by principles common to such classrooms. As NCTM notes, "The Principles for school mathematics reflect basic perspectives on which educators should base decisions that affect school mathematics. These Principles establish a foundation for school mathematics programs by considering the broad issues of equity, curriculum, teaching, learning, assessment, and technology" (NCTM 2000, p. 2).

The Structure

My house of mathematics is not showy, but it is strong. Notice that it resembles a triangle, considered to be the strongest of all basic shapes, with "corners" labeled with the letters *C, I,* and *A*. These letters stand for three major areas of teacher concern: curriculum and mathematics content, instructional pedagogy for teaching mathematics, and assessment. Arrows indicate the interaction of decisions made in each of these fields. I use the triangular model to illustrate the high level of interdependence among content, pedagogy, and assessment in making instructional decisions. I also use it to remind myself that although at times I may focus on mentoring teachers in one area, their success in promoting students' learning in their classrooms is dependent on a deep and connected knowledge base in all three areas.

The "Heart" of the Home

Over the years, I have become increasingly convinced of the need to place children's thinking at the center of our practice as teachers of mathematics. Any decision making regarding curriculum, instruction, and assessment should be based on knowledge of the mathematical thinking of our students and made with the goal of further promoting that thinking.

— *Eula Ewing Monroe*

REFERENCES

National Council of Teachers of Mathematics (NCTM). *Curriculum and Evaluation Standards for School Mathematics.* Reston, Va.: NCTM, 1989.

———. *Professional Standards for Teaching Mathematics.* Reston, Va.: NCTM, 1991.

———. *Assessment Standards for School Mathematics.* Reston, Va.: NCTM, 1995.

———. *Principles and Standards for School Mathematics.* Reston, Va.: NCTM, 2000.

Wolodko, Brenda L., Katherine Willson, and Richard E. Johnson. "Metaphors as a Vehicle for Exploring Preservice Teachers' Perceptions of Mathematics." *Teaching Children Mathematics* 10, no. 4 (December 2003): 224–29.

Learning from Mentors as a Beginning and Experienced Teacher

My experience with mentors both as a new teacher and somewhat later in my career shows that the mentoring relationship can be beneficial for teachers at any stage in the profession.

The Beginning

When I was a first-year teacher, Mr. Alexander was assigned as my mentor. Without his support and professional advice, I do not think I would have been as successful as I have been as a mathematics teacher. My college coursework had prepared me in some ways, but I still lacked a great deal of knowledge about being an effective mathematics teacher in a real classroom. Mr. Alexander taught me more in two and a half years as my mentor than I had learned in four years of undergraduate teacher preparation. He showed me the true meaning of hands-on mathematics teaching and learning—using mathematics manipulatives and centers—and how to build a strong mathematics learning community. After having such a powerful mentoring experience with Mr. Alexander, I had no doubt that I could be a successful teacher in any school.

Returning Home

My next destination was an urban, high-poverty elementary school in Georgia. I was certain the adjust-ment from one classroom to another would be simple, but I soon discovered that I was mistaken. During new teacher orientation, I felt a little insulted when I realized I would have to participate in a district-mandated mentoring program for one year. I thought to myself, "But I'm not a new teacher! I have two and a half years of experience, and I think I know what I'm doing!" The week before school started, however, I learned that my teaching position had been changed from my "comfort zone" of second grade to prekindergarten. When school began, I felt overwhelmed and unsure of my abilities as a teacher at this level. My "rescuer" came in the form of Mrs. Johnson, who noticed my look of uncertainty as soon as she entered my classroom. She first told me not to worry about appearances and instead to focus on possibilities, words of advice that have stayed with me throughout my teaching career. All the resentment I felt about being assigned a mentor quickly vanished. Mrs. Johnson explained the scope and sequence of the prekindergarten mathematics curriculum and provided assistance related to instructional planning and decision making. She and I met regularly, and at each meeting, we established goals and a time frame for meeting those goals. I found this format helpful for planning, and with Mrs. Johnson's guidance, I thrived as a prekindergarten teacher!

— *Shonda Lemons-Smith and Lisa Matthews*

Section 2:
Who a Mentor Is

VISUALIZE a job posting for a mentor of a beginning mathematics teacher. What qualifications should be included? Can anyone who teaches mathematics be a mentor, or should the defining qualities of effective mentors be identified, then sought after in potential applicants to fill the job?

The previous section began to scratch the surface of the complexities of mentoring a teacher of mathematics. As we delve deeper into the role of mentoring, we begin to explore qualities and skills a mentor might possess to have a beneficial and productive impact on beginning mathematics teachers. The nuances of a school's mission, culture, and environment contribute to the specific skills and characteristics desirable in mentors, yet the authors of the articles in this section paint a similar picture of an effective mentor regardless of the school, grade level taught, or professional experience of the mentor.

66 Mentor: someone whose hindsight can become your foresight. 99

—*Unknown*

Will You Be My Mentor?

Will you be my mentor?
Will you be a helpful friend or a judge?
Will you appreciate my enthusiasm and be caught
    up in my excitement or ridicule my youth
    and inexperience?
Will you be my partner in new adventures and
    help me put my great ideas into practice?
Will you guide me with a gentle, experienced
    hand?

Can you travel back in your mind and remember
    what I am feeling and the kind of guidance
    I need?
Can you listen to my problems with an open heart
    and mind?
Can you help me learn from my mistakes?
Can you help me refine my teaching by sharing
    your experience?
Can you make time for me in your busy
    schedule? May I call you in the evening
    when I am worried about how to reach a
    troubled student or interact with an angry
    parent?

Do you really want to be my mentor? Am I a job
    or a professional responsibility?
Do you care about my success as you care about
    the success of your students?
Do you respect me? Do you view me as a peer?
Do you expect to learn and grow from our
    relationship?

Are you the right mentor for me?
Are you the person who will make a difference in
    my life, my teaching, and my career?
Are you the colleague who will help me become
    the best teacher I can be?
Are you the teacher I'll write about one day? "Let
    me tell you about my mentor, my friend."

The questions posed in this poem echo the reflections of novice teachers and mentors gathered during five year-round mentoring workshops in Oklahoma. These workshops[1] were intended to bring together novice teachers with a cadre of supporting mentors who were uninvolved in the first-year evaluation process. For the novice teachers, the workshops established a safety zone: They found a willing group of mentors eager to assist them and guide their progress. For the more experienced participants, the workshops fostered interactions with novice teachers that allowed them to redefine their roles as mentors. These experienced teachers found the role of mentor to be more than a professional obligation. Being a mentor requires the physical, mental, and emotional commitments and contributions of a good friend.

—Gwen Carnes

Five Essential Responsibilities of an Effective Mentor

Mentors play an essential role in the development of new teachers and, to be effective, must carry out certain activities in support of their protégés. The following list of five essential responsibilities of an effective mentor was gathered from discussions with mentor-teachers and protégés.

Providing a Solid Foundation

As the support structure for a new teacher, the mentor assists his or her protégé with day-to-day administrative tasks to allow the protégé to focus on teaching. Moreover, the mentor supports the protégé in handling discipline issues that might hinder effective teaching.

Sharing Ideas and Information as a Good Colleague

A good colleague might be defined as a coworker who takes on the other four responsibilities listed here. As a good colleague, a mentor keeps his or her protégé informed about school events or traditions, shares ideas, and co-plans. True collegiality benefits both the protégé and the mentor. The protégé gains the benefit of the men-

---

1. The workshops were funded through the Oklahoma Teacher Education Collaborative, a project of the National Science Foundation; the Oklahoma State Regents for Higher Education; and the University of Central Oklahoma.

tor's insider knowledge of the school system in which they both work while the mentor learns new ideas from the protégé and may be inspired to vary classroom activities by the process of knowledge sharing.

Acting as the Protégé's Cheerleader

Beginning teachers need praise and encouragement. An effective mentor realizes when his or her protégé is struggling and tries to boost the protégé's morale. The mentor may work with the protégé during planning time or offer encouraging words over a cup of coffee or a meal after school.

Promoting Reflection on Practice

Good mentors encourage reflection on practice in their protégés. Having new teachers reflect on the events that take place in the classroom rather than tell them what went wrong or right helps them develop self-reliance and the skills to become better teachers. In this role, mentors need to strike a balance between providing support and micromanaging their protégés. Encouraging reflection helps new teachers think about and solve problems for themselves, not become mimics of their mentors.

Offering Constructive Criticism

Finally, effective mentors provide timely and relevant feedback. Protégés need encouragement and praise, but they also need to know where they can make improvements in their teaching. If the mentor has provided a firm foundation for the protégé, acted as a good colleague and cheerleader, and encouraged reflection on practice, then the protégé is more likely to be receptive to constructive criticism. If the mentor has not fulfilled these roles, the protégé may become defensive. Receiving feedback that enables learning and improvement is essential to the growth of the protégé. Mentors must learn to provide this feedback in a way that prompts acceptance rather than rejection.

—Jeremy Winters and Jason D. Johnson

Transitioning from Protégé to Mentor

Could I make the transition from protégé to mentor in less than a year? In my third year of teaching and my second year at the same middle school, remarkably enough, I had achieved seniority among the mathematics educators in the school. Something seemed terribly wrong with that picture. Why are schools unable to recruit and retain highly qualified mathematics teachers? I honestly believe that new educators leave the field because of unrealistic expectations and lack of training in the mundane daily tasks associated with being a teacher. Mentors may be one solution to this common problem.

One thing my undergraduate program in education lacked was preparation for what I think of as the "daily grind" of teaching, part of what a good mentoring program should address. Often, new teachers are unaware of the fact that teaching involves much more than just delivering mathematics instruction. Looking back over my teacher preparation, I sometimes think I must have missed signing up for the classes I really needed, including Conflict Resolution with Parents and Colleagues, Students in Crisis: How to Get Students Who Struggle with Daily Basic Needs Invested in Their Education, and Meeting the Demands of Standardized Testing. Only through the efforts of my mentor was I able to keep afloat. During my first year at my current school, I felt like a refugee from the *Titanic*. The only thing that kept me bobbing around in the ocean of middle school was the support and understanding of my mentor. I could say anything to her, bounce my thoughts off her, and ask for advice. She did not judge my ideas, and she made me feel important when she wanted to use some of my activities in her classroom. I think the essential element in developing and maintaining a great mentor-protégé relationship is that both parties believe the relationship is individually rewarding.

Recently, I was asked to be a mentor for a mathematics teacher who outranks me both in age and educational experience. The first words that came out of my mouth were "Are you sure I am qualified to serve in that role?" My life experiences were limited, and teaching was still new to me, as well. The union president, who sets up the mentorship program and is head of our K–12 mathematics department, responded with a look of surprise, exclaiming, "You absolutely are!"

The more I think about the job of mentor, the more I believe that the most important role is to be a personal confidant, someone who can be told anything, from the protégé's deepest fears about teaching to mistakes made in the classroom. I believe that a mentor is someone with whom a protégé can share ideas without feeling foolish and someone who asks the right questions to help other teachers learn and reflect on the experiences they have in the classroom. As I said, having a mentor in my first year of teaching was like being thrown a life raft to keep me afloat on the waves of parent conferences, committee meetings, and paperwork. A good mentor knows when to patch or redirect the life raft or leave it alone to drift but will never let the protégé feel as if he or she is alone in the teaching experience. At times, I still feel that I am lost in the waves, but I now know where and when to turn to someone for help. My role as a mentor is to share this knowledge with my protégé.

My expectations were high when I initially entered the profession as a beginning teacher; I thought that my struggles in the classroom would be limited and that all students would thrive and want to learn. I still have high expectations for my students and their parents, the school administrators, and myself as a teacher, but I know that any of us may be hit by waves or caught by the undertow. My experience has taught me to bring an extra wetsuit for my protégé, and together, we will face the surf.

—*Jennifer Drinkwater*

To My Once and Forever Master Mentor

Dear Jim,

I was thinking the other day that it has been almost forty years since we first met! It seems like only yesterday that I was a twenty-one-year-old student-teacher of high school mathematics, as green as green can be, and you were a veteran of many years, classes, and students; the mathematics department chair then and for my six years at the school; and a professional interested in all

the numerous "_CTM" organizations, whether G for Gary, I for Indiana, or N for National. You certainly knew a lot of mathematics, which you shared with our students—and me. When they (or I) did not understand a new idea, you always had "one more way" to show it. When I did not have a car to get to a "super" mathematics teacher meeting, you provided door-to-door service and back. When I forgot my coat for the annual individual school picture, you shared yours with me and another teacher so that we would "look professional." (Later readers, paging through that yearbook, will wonder if that coat was some kind of 1970s fad.) As a colleague, no matter how busy you were getting ready to teach a class or night school, you always had time to discuss students, the art of teaching students or teaching mathematics, my upcoming marriage, my children, or my plans to return to graduate school for a doctorate. In the dictionary, the word *mentor* should be— and in my mind and heart is—accompanied by your picture.

The years have passed quickly, and I thought that I would share with you some of my personal news: I am still married, I have a great family, and I got that graduate degree in mathematics education. In more than thirty years since I left the old school, I have taught more than 1500 secondary preservice educators, most just as green as I was, with more than 200 of them specializing in mathematics. Of course, during that time, I have also worked with their supervising teachers, and Jim, I think you would be proud to call this fine group of individuals colleagues. They fulfill many of the same roles with their protégés as you did with me.

When my mathematics student-teachers ask me what their student teaching and their supervising teachers or their first year of "real" teaching and their mentors will be like, I tell them that I hope their experiences will be as rewarding as mine was with a master mentor, you.

Always,
Bill

—*William D. Jamski*

Who Should Be a Mentor?

Anyone seeking to become a mentor should reflect on whether they have the interpersonal and professional attributes listed below.

Interpersonal Attributes

- Enthusiasm for the subject and the ability to inspire other teachers
- Belief in the mentee's ability to be successful as a teacher
- Encourages beginning teachers to try new instructional techniques and methods
- Ability to foster trust and respect
- Interest in listening to others, openness to suggestions, and the ability to be fair-minded and sensitive to others' needs

Professional Attributes

- Strong content knowledge
- An effective teacher
- Analytical skills enabling the mentor to find meaningful solutions to the mentee's concerns
- Flexibility to offer a variety of teaching suggestions
- Passion for using new teaching techniques and delivering new content
- Knowledge and desire to monitor and assess the mentee's progress and identify areas for improvement

—Andrzej Sokolowski

Having Influence versus Having Power

The goal of Project Mentor[1] was to develop capacity in small island communities to support mathematics teaching and learning by training local mentors. In turn, those mentors would lead workshop activities for classroom teachers. Over weeklong summer sessions, the mentors participated in activities to increase their own understanding of mathematics content and skills in mentoring. Then, they were expected to conduct similar professional development activities for teachers in their local communities.

Grace was a classroom teacher, the youngest participant in the four-year project. Several other classroom teachers were also involved in the project, as were participants whose positions as curriculum coordinators or mathematics specialists allowed them to regularly engage in professional development opportunities with teachers. Grace anguished over the expectation that she would conduct professional development workshops for more experienced teachers. She was one of the youngest teachers in her small community and was related to many of the other teachers in her school. For her to approach these older teachers with new ideas would be considered culturally inappropriate. As one of the youngest teachers at the school, Grace was expected to learn from the more experienced teachers, not try to pass on her learning to them. She was concerned that she would be perceived as arrogant if she offered a workshop or even took the initiative to approach other teachers with the ideas she learned through the project.

Throughout the summer project meetings, the forty mentors shared various options they had for engaging their colleagues in project activities. Some planned to conduct summer institutes, and others would convene sessions after school. Almost all these approaches involved the mentor's initiative in recruiting local teachers. Grace knew these options for inviting or requiring her colleagues to participate would not work for her. She voiced her concerns to the project leaders because she was sure that she would not have an opportunity to share what she had learned if doing so meant that she had to initiate activities with teachers at her school.

The following summer, when the project participants reconvened, mentors were asked to share the work they had done locally. Surprisingly, Grace had one of the best experiences to share. Although she had no power to organize a workshop with her colleagues, she did have influence through the work she could do with her own students. During the school year, Grace did not approach

---

1. Project Mentor was a four-year program sponsored by the University of Hawaii and the Pacific Resource for Education and Learning.

the other teachers at her school. Instead, she engaged the children in her classroom in the activities she had learned. The students responded enthusiastically and talked to students in others classes about the interesting assignments and activities they were doing in mathematics. When the older teachers heard about these activities, they approached Grace and asked if she would share her ideas. Grace's influence on the practice of other teachers came not through imposition but through their own appeal. Although she did not have the power to impose change, she had the influence to invite it.

—Joseph Zilliox and A. J. Dawson

Section 3: What a Mentor Does

THE first two sections of this book explored why mentoring beginning teachers of mathematics is important and examined characteristics and qualities sought after in someone who mentors. Yet many readers of this book may have skipped the first two sections and turned immediately to this one. Just as classroom teachers do, mentors search for ideas and suggestions on how to improve their craft. In this instance, the craft is mentoring.

As a reader of this book, you recognize the value of mentoring beginning mathematics teachers. Maybe you are already involved with mentoring at some level and still realize that you can, perhaps, do more in your role as mentor. Reading through this section, you will find that a mentor's roles and responsibilities can be varied and far-reaching. The numerous articles that follow will undoubtedly support some existing beliefs you have about mentoring teachers of mathematics while also helping you think about mentoring in new ways.

66 Whoever ceases to be a student has never been a student. 99

—Unknown

The Five-Star Mentor: A First-Class Guide to Teaching

Five-star (five·star: *of first class or quality*[1])

The job of a mentor is to acclimate and empower a mentee with the goal of turning out an independent and successful worker or, in our relationship, a confident and effective teacher. I have been fortunate over the years to learn and grow under some brilliant mentors—influential individuals who were able to encourage but not push, support but not coddle, suggest but not criticize, and lead but not dictate in the context of mathematics education. With this background, I have reflected on my growth under the tutelage of those phenomenal guides in an effort to suggest what I believe to be the cornerstones of effective mentoring. These cornerstones are described in the following paragraphs to enable others to benefit from what I call the five stars of good mentoring.

Investment (in·vest: *to involve or engage*)

Both mentors and mentees must be "invested" in their relationship. The mentor, however, is primarily responsible for developing a shared sense of commitment. Initially, most mentors sketch out tentative plans for their mentees. Such proposals, which may include observation schedules, lesson ideas, and collaborative meeting times, should be discussed at the first meeting and revisited thereafter to ensure that the needs of the mentee are being met. Further, the mentee should have a voice in issues to be addressed. After all, he or she may be nervous about teaching, student relationships, or basic survival in the classroom; thus, the mentor may focus on these aspects of the job early in the relationship. Providing useful, timely support fosters a sense of investment in both parties.

Protection (pro·tect: *to cover or shield from injury, damage, or destruction*)

Mentees must feel safe in their new environment. As I remember, walking into a building full of twenty-year veteran teachers, experienced administrators, and impressionable youngsters can be quite intimidating. Mentors and others must make new teachers feel welcome by inviting them to participate in collaborations, lesson planning, and similar activities. Further, new teachers bring a sense of novelty and creativity to the job. If comfort and safety are established, they may be more willing to take chances in mathematics instruction. In my rookie year, I was fresh from a great teacher preparation program and full of ideas. Rather than sift through each lesson I had planned, my mentor, Noreen[2], encouraged me to try out my ideas: "You can always modify activities in the future," she said. "After all, if you never make a mistake, how do you learn?"

Consistency (con·sis·tent: *marked by harmony, regularity, or steady continuity*)

Being a mentor is difficult. After all, you have your own classroom, your own life, and your own well-being to worry about. The good news is that most mentees understand this reality. They do not expect their mentors to be in the classroom after every class period to answer questions, nor do they expect to receive baskets of engaging teaching materials every week. In fact, a mentee may be so wrapped up in acclimating to his or her new surroundings that excessive attention may be suffocating. The mentor should be available, however, if needed. Before school began, Noreen and I met a few times to discuss questions I had and to have a little bit of fun. She stopped by several times during my first week, offering to sit down and talk if I needed to but not pushing me to confide in her. Occasionally, she would send me an inspirational e-mail or nice note or catch me in the copy room for a chat. Soon, I was off on my own, joining various committees and organizing assorted events. Her steady, reliable support, though, continued to help me immensely.

Longevity (lon·gev·i·ty: *long continuance*)

Investing in your mentee and offering initial protection may seem easy at first, but mentors should remember that these types of support must be lasting. The relationship is not sustained for just a few weeks or months but until the mentee feels comfortable as a teacher. Even

---

1. All definitions are taken from the *Miriam-Webster Online Dictionary* (http://www.m-w.com).

2. All names in the article are pseudonyms.

then, he or she may come to you with questions or suggestions or just to talk, and your support at these times must be unwavering. Of course, all teachers have busy weeks, and experiencing a few days of minimal communication does not mean that all is lost in the relationship. For mentees, just knowing their mentors are available for the long haul helps get them through the day. From September to May, Noreen observed my class; assigned me to observe others; and checked on me through e-mail, informal meetings, and notes. She backed off slowly, letting me get my feet wet, but somehow, I always knew she would be available for me. This relationship continued during my tenure at the school, and although years have passed and we have both moved on to other educational settings, we still correspond through e-mail. Our relationship has passed the test of longevity.

Extension (ex·ten·sion: *an enlargement in scope or operation*)

The relationship between mentor and mentee should also be extended beyond school. An occasional cup of coffee or dinner out may encourage both parties to open up and discuss important topics on their minds. Being physically removed from the school setting allows both mentors and mentees to "let loose" while still discussing school issues. Such occasions may also offer the opportunity for mentees to meet other school employees who may work in different departments or locations, broadening the support and social circle for new teachers. Mentees will greatly appreciate any efforts you make to help them relax and have some fun.

Indulging your mentee in luxurious, five-star treatment is sure to produce first-class results in a new teacher.

—*Emily Peterek*

What I Learned from My Mentor: Supporting Beginning Mathematics Teachers

In my position, I mentor beginning mathematics teachers each year, and each year, I reflect on the qualities and responsibilities of a good mentor. As I think about effective mentoring, my thoughts often return to my mentor and my first year of teaching. Ellen (a pseudonym) was a seasoned veteran of twenty-five years, and she loved teaching. Through her mentoring, I started to make a contribution to my students as a mathematics teacher and learned about my identity as a teacher. Ellen was my "official" mentor, but much of the guidance she gave me was informal. Mentoring is no easy task, but Ellen made it seem effortless and comfortable, and our relationship was an open and supportive one. She inspired me to strive always to be better in the next lesson and to get to know my students as people, not just learners of mathematics. She always valued my ideas but challenged them in the spirit of debate that was indicative of her passion for teaching students mathematics.

I learned many things about teaching mathematics from Ellen, but my focus here is to identify her top three strategies for developing an effective mentoring relationship. I have tried to use these same strategies through various phases of my life, and they are invaluable to me as I interact with beginning teachers today. In my mind, the most important three strategies used by good mentors are to value beginning teachers' ideas, discuss the reasons behind instructional decisions, and provide an alternative lens through which the new teacher can view the classroom experience. I offer a glimpse of each strategy in the following paragraphs.

Value Beginning Teachers' Ideas

Ellen always asked my thoughts on activities, students, and teaching. Following one observation session, she began our formal meeting with the question "What did you think of the lesson? What do you think students learned about mathematics?" I talked at length about the aspects of the lesson I found useful and those I might change in the next iteration. She encouraged me to share my thinking, concerns, and questions with her, and she took time to consider my comments carefully. She treated me as a colleague, and both her gestures and comments emphasized the importance of my ideas. This seemingly simple act of respecting my thoughts had significant implications for my relationship with Ellen and my view of myself as a teacher. I felt that Ellen was invested in me and my teaching, and I felt competent in the classroom. She believed in me and conveyed a genuine interest in my thoughts on teaching, even though she was clearly the expert, with valuable insights in most situations.

Discuss the Reasons behind Instructional Decisions

Ellen consistently initiated conversations with me about the reasoning behind instructional decisions. I learned a great deal from these conversations because they fostered insights into the decisions she made before, during, and after mathematics lessons. Her "suggestions" to me were not presented as lists of do's or don'ts. Rather, they surfaced through everyday conversations about instructional decision making. Through this practice, she also modeled reflection and an outlook that continuously examined instructional decisions in light of students' learning. Ellen emphasized the importance of understanding the reasons behind teachers' instructional practices, not only the effective or "best" practice itself. In retrospect, I believe this understanding offers beginning teachers a toolbox of practices to work with and an understanding of why each one may be useful. They are then able to draw on and adapt practices to work in their specific classroom situations.

Provide an Alternative Lens

Often during our mentoring relationship, Ellen suggested different ways to look at classroom situations to give me insights that I had not yet discovered on my own. In this way, she provided an alternative lens through which to view events in the classroom. If I talked with her about a particular student's difficulties in learning a concept, she might offer a number of possible explanations for the situation. She might ask about the student's home life or about specifics of how the student solved particular problems. In doing so, she fostered the idea that many factors may affect classroom performance. Ellen rarely stated an answer absolutely; instead, she indicated potential responses. In this way, she helped me realize that teaching involves analysis and refinement and that it is rarely if ever perfect. In fact, her motto seemed to be that one of the most interesting aspects of teaching is that it can always be improved. Teachers can always do better in the next lesson.

Results

Through Ellen's support and expertise, I came to view reflection and change as natural parts of teaching. For her, mentoring was about open communica-

tion, trust, and mutual respect. She never pushed me to change, but she encouraged me to view change positively and to carefully and intellectually consider changes in my practice that would enhance student learning and interest in mathematics. As we worked together, Ellen came to understand the teacher I wanted to become, and she respected my choices. She supported me in fulfilling my goals as a teacher instead of trying to make me a carbon copy of herself. Her task was not a simple one, but in my view, the strategies she used lie at the heart of effective, supportive, and thoughtful mentoring of beginning teachers.

—Lynn Liao Hodge

Adding to the Repertoire: How Mentors View Their Roles

During an intensive, three-year leadership development project, we asked thirty mentors each year how they viewed their mentoring roles. Every year, the mentors conducted twenty-one hours of seminars for beginning teachers, visited classrooms, taught model lessons, and coached beginning teachers; they were supported in these endeavors by participation in annual institutes and monthly leadership development meetings throughout the three-year period. As we analyzed the mentors' responses after each data-collection cycle, we found that the mentors played three general roles: resource, relationship builder, and change agent.

The Mentor as Resource

At the beginning of the project, mentors generally envisioned their role as providers of information and teaching strategies for beginning teachers. Mentors saw themselves as resources for ideas, materials, mathematical knowledge, and teaching tips. One mentor wrote, "I see myself as a resource to those teachers [whom] I will work with. My goal is to identify the individual needs of the beginning teachers. I want the teachers that I work with to feel positive about themselves and their math instruction. I hope my support will ease some of the frustration many of our new teachers feel about teaching math."

The Mentor as Relationship Builder

After the first year in the project, some mentors began to see the importance of building a trusting relationship with beginning teachers. Building a relationship meant listening, helping to create a safe environment for mathematics learning, and allowing teachers to communicate openly about their fears and successes in the classroom. Again, one mentor expressed a personal view of this role:

> During this year, I've gained a lot of confidence as a mentor. I have a growing understanding that my relationships are central and the most useful part of my program with the beginning teachers. As they begin to trust me, they ask questions more freely, and they begin telling me about strategies they tried and how they succeeded or failed. I've done a great deal more listening than I expected, and I've also learned much from the teachers I initially set out to help.

The Mentor as Change Agent

We define a *change agent* as someone who takes responsibility for asking questions, introducing topics of equity and access, and bringing to the forefront issues that are sometimes ignored at the district, school, or classroom level. *Principles and Standards for School Mathematics* (NCTM 2000) states that equity is fundamental to high-quality mathematics education and, therefore, must be an essential element in developing teacher-leaders in mathematics. Over a three-year period, the project incorporated many opportunities for mentors to work collaboratively to define and explore equity issues. Generally, as mentors felt more confident in their roles, they brought up equity issues with beginning teachers more often. Some mentors in their second and third years wrote about their roles as change agents:

> I have always seen myself as a worker for justice, but up until a few years ago, I saw very little connection between equity and math. I now see increasing ways that equity affects student learning opportunities, especially in ways that are unspoken. I see my role expanding in leading students to see themselves as strong and capable, not limited by gender or color or class. And I am more willing to question, to speak up, and to stir up the comfortable acceptance of "the way things are."

> I see myself as an advocate for students who may not learn math in traditional ways. By helping to facilitate seminars for beginning teachers, I provide new teachers a chance to see and experience many different strategies for doing mathematics. We all experience and learn math differently. I view my role more as an agent of change.

Conclusion

Although we have documented three roles that mentors played in the mentoring process, we do not mean to suggest that mentors should shed one role to adopt another. Rather, mentors accumulate roles over time, continually adding to their repertoire.

We invite you to reflect on your responsibilities as a mentor and the various roles you have incorporated into your mentoring. We hope this project summary serves as a prompt for mentors to reflect on their multidimensional mentoring practice and see the importance of their work, not only in their own professional growth but also in supporting newcomers to become effective teachers of mathematics.

—Nancy O'Rode and Nancy Terman

REFERENCE

National Council of Teachers of Mathematics (NCTM). *Principles and Standards for School Mathematics.* Reston, Va.: NCTM, 2000.

Mentoring at Any Age Is Important

For me, the top five benefits of having a mentor in my teaching career are as follows:

- My mentor provided a place to turn for all my questions.

- My mentor benefited from my enthusiasm.

- Having a mentor meant that I knew at least one person in the school building.

- Knowing I had a colleague to turn to for support boosted my confidence.

- My mentor kept our conversations confidential.

—Ann M. Perry

Mentoring Mindsets for the Mathematics Teacher

Experienced teachers understand that no one "right" way exists to teach mathematics, but having the "right" attitude is important for any teacher. Having the right attitude means maintaining a healthy mindset when teaching mathematics and sharing ideas about teaching with in-service and preservice colleagues. This article describes some essential aspects of a mentor's mindset when coaching in-service or preservice teachers on effective skills for mathematics instruction. In particular, we discuss the strategies of speaking mathematically during instruction, anticipating students' questions or misconceptions, and using a variety of examples to represent mathematical concepts.

Speaking Mathematically during Instruction

Experienced teachers generally realize that speaking mathematically during instruction means more than just telling students what to do. According to Bratina and Lipkin (2003), "teachers should model communications that are precise" (p. 11) in conveying mathematical ideas during instruction. Beginning teachers should make sure to use appropriate terminology and phraseology when teaching and should require the same behavior from their students. For example, in talking about a door, a new teacher may call the shape a *square* rather than a *rectangle;* if the teacher misuses terms frequently, students, too, may fall into the habit of using such terms incorrectly.

At times, new teachers may also omit necessary words in phrases or fail to clarify certain phraseology. For instance, a fraction is often defined as "a part of a whole." This definition may be appropriate for younger students, but it may lead to confusion in the upper elementary and middle grades when the definition of a fraction as "a ratio" is more appropriate. When discussing probability, a protégé might say, "The probability of landing on heads in a coin toss is one-half" without clarifying that this result is theoretical and that the probability may vary under experimental conditions.

Anticipating Students' Questions or Misconceptions

Anticipating students' questions or areas in which students may have difficulty can make instruction more effective, but foreseeing potential problematic areas in mathematics may not always be easy for beginning teachers. To address this weakness, the protégé and mentor may plan lessons collaboratively, thus helping the protégé organize more comprehensive lessons, prepare effective questions, and predict possible student errors. For example, a mentor-teacher explained to her protégé that second-year-algebra students often experience difficulty in translating between exponential and logarithmic forms of equations. Sharing this information with the protégé enabled the two to discuss a plan of action to reduce student confusion.

For the mentor, helping the protégé anticipate students' questions assists in connecting the "big ideas" in the mathematics he or she teaches. For instance, experienced teachers often note that students perform better in learning the four basic arithmetic operations when they have a firm understanding of place value. Mentor-teachers can help their protégés discover ways to teach the basic operations that enhance students' understanding of this important concept.

Using a Variety of Cases to Represent Mathematical Concepts

Finally, experienced teachers often teach by using examples or introducing situations that require students to use inductive or deductive reasoning skills. Mentor-teachers know the importance of using a variety of examples or cases when applying inductive reasoning; however, new teachers may not be aware that, for example, they habitually use squares to signify quadrilaterals, which could lead students to associate only squares with quadrilaterals.

Ultimately, we believe that one of the most important tasks in teaching mathematics is to foster students' ability to use spatial visualization skills to construct their knowledge. Mentors may realize that students have trouble with graphs when the axes do not have a scale of 1, but new teachers may always use graphs with x and y

scales of 1. When students are given examples to graph that call for large numbers, they may have difficulty plotting points because, on the paper they are given, they cannot fit a graph that extends that long. To address this misconception, new teachers should show examples of graphs that use a variety of x and y scales appropriate to the data given.

Conclusion

The mentor-teacher's outlook on the teaching and learning of mathematics is vital to the development of their protégé colleagues. The three mindsets emphasized in this article are essential for empowering mentor-teachers to develop supportive dispositions to help beginning teachers establish long-term careers in teaching mathematics.

—*Jason D. Johnson and Michaele F. Chappell*

REFERENCE

Bratina, Tuiren A., and Leonard J. Lipkin. "Watch Your Language! Recommendations to Help Students Communicate Mathematically." *Reading Improvement* 40, no. 1 (Spring 2003): 3–12.

"Aiding and Abetting" Teachers of Mathematics

The phrase *aiding and abetting* may conjure up less than noble associations; however, when the phrase is examined for closer meaning, one discovers that *to aid* means "to help or assist" and *to abet* means "to encourage or support." This article examines the phrase *aiding and abetting* in the context of education and, more specifically, the context of mentoring teachers of mathematics.

As a mentor to teachers of mathematics, I see my role as aiding and abetting new teachers in building confidence and developing content knowledge. For many of the teachers I mentor, our interaction takes place through e-mail correspondence or telephone conversations.

Teachers I mentor commonly ask, "What do I do when …?" Usually the situation that comes after "when" involves a student's misunderstanding of a concept or a procedure. For example, one new teacher recently asked, "What do I do when my students confuse area and perimeter?" To aid and abet in these instances, I often begin by trying to understand why the mentee needs to pose such a question. My goal is not necessarily to answer the question but to empower the mentee to construct his or her own appropriate answer. In this instance, I asked the mentee, "Are the students confusing the concepts of area and perimeter or the formulas for area and perimeter?" Sometimes when teachers focus solely on the formulas using the typical variables of l and w (area = $l \times w$ and perimeter = $2l + 2w$), students can become confused about the formulas. To address this problem, the teacher can engage students in learning activities designed to help them build their understanding of the concepts, even to the point of generalizing formulas in a way that does not use the traditional variables l and w. By helping the mentee "step inside the mathematics" of area and perimeter as concepts rather than formulas, I aimed to help her develop strategies that would prevent such confusion among learners.

As you can see, I am guilty of aiding and abetting the mentees assigned to me—assisting and supporting their work as teachers of mathematics. I acknowledge this example as evidence of my guilt, and I accept my life sentence of commitment to mathematics teacher education!

—*Thomasenia Lott Adams*

A Powerful Partnership

A mentoring relationship is a powerful partnership, with each party benefiting from the other's contribution. From the perspective of a mentee, a good mentor serves as a resource and friend, influencing a beginning teacher in the following ways:

- Helping you see how daily lessons fit into the big conceptual picture of mathematics

- Congratulating you for creating lessons that teach concepts in deeply meaningful ways

- Correcting conceptual errors that you may inadvertently perpetuate among your students

- Offering a pep talk after your first parent conference

- Seeing you several times a week at school

- Appreciating that you have a life away from school and being genuinely interested in that life and how it influences your teaching

- Maintaining confidentiality

- Helping you find practical approaches that stay true to your teacher education training and are effective at the classroom level

- Understanding that the education field has changed since he or she went to college and showing eagerness to learn about fresh perspectives and current research-based approaches

- Showing you how to use the copy machine, where supplies are stored, and how to navigate the school's automated substitute-teacher scheduling system

- Not claiming to have all the answers

- Wanting you to figure out some answers on your own

- Asking for your opinion about his or her own teaching challenges

- Wanting you to succeed

—Carrie S. Cutler

"I Don't Need to Have All the Answers"

One of the best discussions we had at our monthly mentors' meeting was the one that began with one mentor's realization "I don't need to have all the answers."

The mentor was relating the unease she felt in leading her group of new teachers, knowing that she did not have all the right answers to their questions. After several months, she realized that she did not need to have all the answers; what she needed was to trust her beginning teachers enough to know that one of them would have an idea to solve the problem under discussion. She then began to work on building trust in the group of new teachers so that everyone felt free to ask—and answer—questions.

As soon as this mentor was finished speaking, the others applauded her idea. Now our group happily shares the burden of finding answers with their mentees.

—Nancy O'Rode

So What Do You Want from Me?

The first few weeks of school are always tiring for me. For many years, I thought that the source of my trouble was the transition from the lazy days of summer to the craziness of the school year. As I grew into a more experienced teacher, however, I came to understand that those weeks represented a different challenge. I spent the first few weeks of the school year getting to know my student population and figuring out what each of the students needed from me to be successful. Does this student need a pat on the back or a little push? What can I do to help this student be more successful?

In the same way, when I started working with teachers, I discovered that my credentials and experience mattered less than the guidance I could offer to meet the needs of these adult learners. Student-teachers and teachers in their first few years of practice are faced with significant challenges. Like students, they represent a variety of learners with a variety of needs. When I figured out how to meet some of those needs, I became both valuable to, and valued by, these beginning teachers. When they came to me with questions that I was able to answer, my knowledge and experience were appreciated, but if I tried to give unsolicited advice, my words were less welcome. Although my responsibility was to guide them, the teachers had to feel the need for guidance first.

New teachers, like students, require a caring individual who values their contributions as developing professionals. Their successes are as varied as their backgrounds and life challenges. My support for them was only as effective as my discernment of what they needed from me.

—Mary Belisle

My Mentors: Qualities That Made Them Special

Because of the influence of my mentors, I profess the love of mathematics education to all those I meet. I credit my mentors with the fact that I have spent more than ten years teaching mathematics to children and that I do not know any bounds when teaching mathematics. Both mentors should feel proud of sharing a special part of themselves with me and guiding me to become the educator that I am today.

Both my mentors helped develop my professional teaching career. One was assigned to me under a state program that required every first-year teacher to have a mentor. The other chose me as a protégé because we had similar teaching philosophies. Both took the role of serving as my mentor as a serious responsibility. From my experience with these two mentors, I believe that the five attributes described below make a mentor relationship successful.

- Trust is a cornerstone of the relationship between mentor and mentee. Because I trusted both of my mentors, I told them about my classroom adventures willingly and without fear of punishment or reprimand. Beginning and, often, veteran teachers make mistakes in the heat of the classroom moment that we quickly regret. A mentor must be trusted to maintain confidentiality about such "lessons learned."

- Both of my mentors also exhibited positive attitudes. If I believed my lesson plans had not lived up to my expectations, they helped me implement activities and approaches that worked for me. In other words, they found the silver lining in my cloud of inexperience. They provided frequent encouragement when I needed it—sometimes on a daily basis.

- Both of my mentors had the knowledge and experience to offer suggestions for improvement in all aspects of classroom management, from discipline to techniques for teaching certain mathematics topics. The mentor should brainstorm with the beginning teacher to find ideas that work with his or her teaching style to ensure that the beginning teacher feels comfortable with the suggestions.

- Both of my mentors also exhibited an attribute that is difficult to find in today's world: the willingness to help. The mentoring relationship requires both this willingness and a significant time commitment from the mentor to the beginning teacher. This commitment may vary from an hour or two once a week to once a month and should be guided by the needs of the beginning teacher.

- The final attribute of a good mentor is belief in our profession. Many might think that any mathematics teacher automatically believes in the profession, but I would argue that this sense of commitment goes beyond the students. Good mentors realize that as teachers, we have a responsibility to provide guidance and support to newcomers in the field. I was lucky to find two mathematics teachers who believe in our profession and were willing to share that philosophy and their support with me.

—Ann M. Perry

The Mentor as a "Fellow Worker"

The following description of the qualities of successful mentors is based on the model of a "fellow worker" from the findings of Sparrow (2000). Important features of the fellow-worker model are choice, experimentation, and reflection. Principles based on these features are presented below.

The mentor-teacher should have experience working in the context of his or her mentee. The person acting as mentor must understand the context, expectations, and nuances of the environment in which his or her mentee works. For example, a secondary school teacher-mentor may have difficulty guiding or understanding the needs of a teacher who specializes in early childhood education.

The mentee should decide the problem or issue that will be the focus of attention. This principle empowers the mentee and removes the possibility of irrelevance or the imposition of the mentor's agenda. When the discussion is restricted to topics of interest and importance to the new teacher, any suggestions for action will be immediately applicable to the teacher's classroom.

Options for action should be generated through discussions between the mentor and the mentee. This principle embodies one of the main roles of the mentor, that is, to provide a range of possible approaches to address the problem or issue at hand. Each option should be analyzed in terms of its positive and negative aspects and how it fits into the general plan the mentee wishes to follow.

The mentee must decide which option will be implemented in the classroom or elsewhere to address difficulties that arise. This principle also empowers the mentee in that it enables a personal decision about what actions he or she can take given the constraints of the classroom, school, and his or her private life.

Mentors should encourage experimentation in the classroom to generate data and evidence for reflection. Having real data to analyze is essential to promoting reflection on classroom events. Such data are not derived from a book of theory but involve the mentee's students and teaching context.

Mentors should emphasize the importance of reflection on classroom practice and beliefs. The role of the mentor here is to highlight the positive outcomes of classroom experiments because these experiences are more likely to develop positive attitudes in the mentee. These experiences will then be repeated as the beginning teacher grows into the profession.

Professional development should be the "natural process of growth in which a teacher gradually acquires confidence, gains new perspectives, increases knowledge, discovers new methods, and takes on new roles" (Jaworski 1993, pp. 10–11). Teacher professional development through mentoring is based on the open sharing of ideas to give new teachers opportunities to reason about, and learn from, their own teaching experiences.

—Len Sparrow and Sandra Frid

REFERENCES

Jaworski, Barbara. "The Professional Development of Teachers: The Potential of Critical Reflection." *British Journal of Inservice Education* 19, no. 3 (1993): 37–42.

Sparrow, Robert Leonard. "The Professional Development of Beginning Teachers of Primary Mathematics." Ph.D. diss., Edith Cowan University, 2000.

The Chair Incident: An Opportunity Taken

My operational definition of a good mentor-mentee relationship is one in which an experienced teacher and a novice teacher work together to improve students' learning. The seasoned teacher conveys the wisdom gained through reflection on past experiences, and the novice contributes the energy and fresh ideas of one embarking on a new adventure.

Two essential components of mentoring are mutual trust and respect. Mentors should look for occasions to cultivate this type of rapport and take advantage of opportunities to develop trust and respect even if those opportunities are somewhat unconventional. The following paragraphs describe one such opportunity. First, the event is told from the point of view of the mentor, Christie. Next, the mentee, Jessica, describes the event from her perspective.

The Mentor's Story

Jessica was new to teaching, and I was delighted that our district had found a highly qualified teacher to handle the eighth-grade mathematics classes in our middle school. I visited Jessica the day before students arrived to introduce myself and explain my role in the district. She was excited to start teaching but was also disappointed at being assigned to a classroom different from the one she had been shown the day she was interviewed for the position. Her assigned classroom was very small and did not have an adequate number of chairs to seat all students. Before leaving, I told Jessica that I would remind the principal of this problem and assured her that chairs would be found so that her students would not have to stand or sit on the floor in class.

Early the next morning, I stopped by her classroom and discovered that she was still ten chairs short of having enough seats for her students. Being a new teacher, Jessica did not feel comfortable approaching the principal again with this problem. She did not know the custodial staff, nor did she know the building well enough to search out chairs on her own. With only thirty minutes left before students would begin to arrive at her classroom door, I decided to take action myself. I had taught in that building for years and knew every cranny and crevice where chairs could be hiding. My hunt began, and I personally carried chairs to her classroom until she had a sufficient number to seat all her students.

The Mentee's Story

My first year of teaching was interesting to say the least. My classroom was an old janitor's closet that had a makeshift wall separating it into two small rooms. I occupied one half of the room, and another eighth-grade mathematics teacher had the other half. I remember having absolutely no technology or equipment to work with and not even enough chairs to seat all my students. As I was pondering where to seat my students on the first day of school, Christie, my mentor, came by. Christie became my own miracle worker when she found enough chairs for my students that morning. Later, she also found materials and technology for my classroom and showed me new and exciting approaches to teaching mathematics content.

Happily Ever After

Helping Jessica solve this problem initiated the development of a relationship of mutual trust and respect and opened the door for future conversations about other ways Christie could assist Jessica as she embarked on the lifelong journey of becoming an effective teacher. Today, Jessica is a confident and experienced teacher who is often called on to mentor others.

—*Christie A. Perry and Jessica Bacca*

Understanding the Dilemmas of Beginning Mathematics Teachers: Practical Research Findings

In their preservice education programs, beginning teachers are exposed to new directions in mathematics education, including insights from research into how children learn, changing perceptions of what it means to know and do mathematics, the availability of calculators and other technology, the changing needs of society, and the anticipated needs of the workforce in the twenty-first century. When these teachers begin, they often learn that many other factors influence day-to-day classroom events and the teaching and curricular decisions they make in their professional roles. *Dilemmas* arise, that is, situations that cause "beginning teachers to make a decision between two equally important choices" (Sparrow 2000, p. 289). Each choice of action sacrifices possible advantages of the alternative; hence, finding a "perfect" solution is impossible.

What are some of the dilemmas encountered by beginning teachers in the mathematics classroom, and more important, how can a mentor support novice teachers in working through these dilemmas? The following examples are findings from case studies of beginning teachers (Sparrow and Frid 2001, 2002).

Four Basic Dilemmas

1. Choosing between personal beliefs about teaching and the recommendations of others.

   - Beginning teachers are often eager to implement new ideas, such as incorporating mathematical discourse, concrete materials, or group work, in their classrooms, but at the same time, they want to fit in at the school and may feel constrained by their perceptions of the expectations of others.

   - Beginning teachers may feel uncertain about what their students are learning in an open-ended activity in comparison with individual, textbook-based exercises for which students provide right or wrong answers.

2. Selecting and using teaching strategies that focus on developing learners' understanding versus developing their performance.

   - New teachers sometimes feel pressured to "cover" the syllabus or certain pages in a textbook within a specific time frame instead of taking the time to ensure that their students understand concepts and processes.

   - Some beginning teachers may believe that their responsibility is to teach facts and skills in a way that students can remember instead of encouraging students to develop some of their own ways of thinking.

3. Taking risks by trying different teaching strategies versus playing safe and maintaining the status quo.

- New teachers may be unsure about implementing learner-focused or open-ended activities because they lack mathematical knowledge. The alternative is to use worksheets or other methods that will "work" because the correct answers are known.

- Beginning teachers may be reluctant to step outside a comfort zone because they fear lesson "failure." Instead, they may use traditional teaching methods aimed at particular pedagogical goals.

4. Concentrating on the less able students instead of accommodating the diverse range of students in a class.

- New teachers may focus on the middle-level achievers rather than the low or high achievers to maximize the number of students targeted.

- Many beginning teachers feel dissatisfied with "treating all children the same" but are unsure about what resources or approaches they can use to accommodate a greater number of students.

The Mentor's Role

A mentor supports the beginning teacher in the process of reflection, consideration of options related to teaching practices, and small-scale experimentation with options. In this way, the mentor becomes a hybrid being—a trusted friend, knowledgeable resource, colleague, listener, and supportive sponsor. At the same time, a mentor serves as a catalyst to prompt the mentee to examine and reflect on beliefs, values, teaching practices, and school experiences; reason about, and learn from, his or her experiences; and make small-scale plans to implement ideas generated as a result of reflection and analysis.

To ensure empowerment of the beginning teacher, the mentor must generate choices for his or her mentee while avoiding the temptation to reinforce the mentor's own image of mathematics teaching. That is, the mentor must be an "option provider" rather than a "solution provider," so that the beginning teacher engages with and solves problems independently rather than reproduces

the mentor's ideas and strategies for mathematics teaching. Only then can empowerment be achieved.

—*Sandra Frid and Len Sparrow*

REFERENCES

Sparrow, Robert Leonard. "The Professional Development of Beginning Teachers of Primary Mathematics." Ph.D. diss., Edith Cowan University, 2000.

Sparrow, Len, and Sandra Frid. "Dilemmas of Beginning Teachers of Primary Mathematics." In *Numeracy and Beyond,* edited by Janette M. Bobis, Bob Perry, and Michael Mitchelmore, pp. 451–58. Vol. II, *Proceedings of the 24th Annual Conference of the Mathematics Education Research Group of Australasia* (MERGA). Sydney: MERGA, 2001.

———. "Supporting Beginning Primary Mathematics Teachers through a 'Fellow Worker' Professional Development Model." In *Mathematics Education in the South Pacific,* edited by Bill Barton, Kathryn C. Irwin, Maxine Pfannkuch, and Michael O. J. Thomas, pp. 71–80. Vol. I, *Proceedings of the 25th Annual Conference of the Mathematics Education Research Group of Australasia.* Auckland: MERGA, 2002.

Like lesson planning, being a mentor requires a good deal of time and preparation. For example, when is the ideal time to make the first contact with your protégé? With so much information to pass on to your protégé, where should you start, and how much information should you share without fear of overwhelming the new teacher? How soon after the start of the school year should you observe your protégé, and on what aspects of teaching should you focus the observation? A mentor is likely to have as many if not more questions about his or her responsibilities than the protégé has for the mentor.

The purpose of this book is to be a quick and accessible resource for the mentor of beginning teachers. This section in particular furnishes numerous tools to consider, adapt, and use. Several of the articles are designed specifically for the mentor to use with the beginning teacher. For example, "A Guide for Reflecting on Mathematics Lessons with Beginning Teachers" is a user-ready tool a mentor might use to focus discussion about a particular lesson. Other articles, such as "Talking about Teaching: A Strategy for Engaging Teachers in Conversations about Their Practice," offer mentors ideas and suggestions to help develop their own skills in supporting their protégés.

Just as the instructional ideas, activities, and lessons borrowed from colleagues are adapted to fit the specific needs, goals, and skills in individual classrooms, these tools are intended to be modified to suit the relationship of individual mentors and protégés. Perhaps the best approach is to peruse the articles in this section and choose one or two that seem to match your needs as a mentor, adjusting the ideas where necessary to match your goals.

"No man is wise enough by himself."

—*Titus Maccius Plautus*

Developing Effective Mentoring Skills for Mathematics Coaches

Mathematics coaches often face obstacles in their efforts to mentor classroom teachers. For example, when coaches lead a lesson, some teachers might view the activity as designed for the sole benefit of the students, not for themselves, or purely as an enrichment activity. As a result, they may remain uninvolved with the lesson, turning their attention to grading papers or other tasks. Other teachers are not confident in their teaching practices or knowledge of mathematics and resist involvement in coaching for that reason. These conditions are not conducive to effecting long-term change in the teaching and learning of mathematics. To address these concerns, a group of twenty mathematics coaches who served elementary and middle schools in a large urban district developed the following list of strategies for effective mentoring. These strategies focus on the collaborative planning of a mathematics lesson and on postlesson debriefing.

Suggestions for Planning a Lesson with a Teacher

- Discuss the context and climate of the classroom with the teacher to ensure that the lesson fits the teacher's style. For instance, does the teacher feel comfortable introducing a specific activity? Does the teacher use manipulatives often? Do the students work in groups on a regular basis? Have the students had much experience writing about mathematics? Answering these questions helps the teacher and coach work together to devise a plan that best fits the particular classroom.

- Invite the teacher to establish the focus of the lesson by asking, "What aspect of the students' learning do you want to know more about?" The teacher might want to focus on strategies to involve more students in the large-group lesson, to examine the ways in which students justify their reasoning or record their ideas, or to find out how English-language learners exhibit their understanding. Coaches may need to offer suggestions for a focus, but the teacher must make the final decision. In this way, the teacher develops a sense of ownership of the lesson. After the focus is decided, the coach may lead the lesson while the teacher gathers data that pertain to the question to be answered, or these roles may be reversed.

- Discuss with the teacher the reasons behind some of the mathematical decisions in the lesson. For instance, a coach might say, "After working with the students to find all possible ways to use factoring to count on to 10, I would like to challenge them with 6. I picked 6 because it, too, has four factors. I think that some students may be surprised that a number less than 10 has the same number of factors as 10. I want them to confront this result and suggest some theories about why it is true." Sharing this kind of thinking and reasoning helps the teacher understand the coach's instructional intentions.

- Capitalize on the teacher's knowledge of the students' abilities in determining the design of the lesson. How does the lesson meet the needs of all learners? Is the pace of the lesson appropriate? Which students need extra encouragement and support? This approach allows both the coach and the teacher to contribute mutually valuable insights that help ensure the success of the lesson.

Suggestions for a Postlesson Conference

- After the lesson has been conducted, hold a conference in which both the coach and the teacher share "three pluses and a wish." The three pluses are observations about aspects of the lesson that seemed successful. Either the coach or the teacher might refer to an interesting comment or personal connection made by a student, an intriguing question, an unexpected response, or another reaction from the class. The wish involves a plan for the future: How might this initial experience be extended? What changes should be made the next time the lesson is taught? What interesting features of the lesson should be supported in future lessons?

- Be honest and willing to discuss aspects of the lesson that might be handled differently next time. The coach in particular should point out moments during the lesson when he or she felt unsure. By sharing his or her own vulnerability, the coach builds a trusting relationship with the teacher. When coaches demonstrate a reflective stance and pose questions about their own teaching decisions, classroom teachers become more willing to share their questions and doubts.

- Discuss the focus of the lesson, perhaps using the three-pluses-and-a-wish format to frame the observations. If the focus was to challenge students to justify their reasoning, some topics to consider might include their abilities to give evidence, connect several mathematical ideas, note patterns, and make generalizations.

- Provide evidence to support observations, and encourage the teacher to do the same. For instance, if the teacher says, "The students seemed to have a good understanding of the pattern," the coach might ask, "What are some things the students said or wrote that helped show their understanding?" Encouraging each other to be specific helps the two members of the team analyze the experience comprehensively and better assess the students' competence.

- Discuss ways to link lessons over time to build students' understanding. Talk about how this particular lesson might be extended to enable the teacher to pursue related ideas in the future. The lesson should be seen not as an isolated event but as part of a larger plan for the class.

- Reflect on the collaboration process together. What aspects of the planning, teaching, and debriefing processes went well? How is the process of learning together as a coach and teacher similar to the kind of learning we want to foster in students?

Conclusion

These suggestions highlight the importance of a collaboration between teachers and coaches in which each participant shares resources and knowledge that contribute to the design of the lesson.

—Phyllis Whitin and David Whitin

Observing a New Teacher

Of course, mentor-teachers contribute to the professional development of preservice and in-service teachers, but how does a new mentor-teacher know what aspects of teaching to focus on when observing a protégé during instruction? From my experience working with mentor-teachers and protégés, I have noticed some essential elements that mentors should look for during protégé observation sessions. These elements are described in the brief paragraphs that follow. Highlighting these issues in a postobservation meeting will help guide the protégé in developing the skills to become an effective mathematics teacher. Mentors do not have to identify all these items during the course of observing one lesson. Rather, they may assess and discuss various elements over a series of several observation sessions.

Allowing for Adequate Transitions between Activities

The mentor should encourage the protégé to think about the transitions between various activities planned during instruction. For example, the teacher may need to shift the students' attention from group work to a whole-class discussion. The mentor may suggest strategies for making smooth transitions during the lesson.

Eliminating Distracting Verbal or Nonverbal Language

For the most part, protégés do not use distracting verbal or nonverbal language intentionally; however, the mentor may have to tactfully identify distracting behaviors to allow the protégé to control them. The protégé can work to eliminate unwanted behaviors after he or she becomes aware of them.

Fostering a Safe Classroom Environment

The classroom environment is also known as the *classroom culture.* All students should feel safe and free to ask and answer questions in this setting. The mentor should notify the protégé if the classroom culture is not suitable for supportive learning, and offer suggestions that will allow more openness among students.

Asking Questions That Require Students to Think

Protégés may not be aware that the questions they ask during instruction do not require students to use higher-order thinking skills. To remedy this problem, the mentor may wish to help the protégé plan lessons that include questions to promote higher-order thinking.

Emphasizing Student Talk over Teacher Talk

New teachers of mathematics sometimes feel that they must project the image of gatekeepers of knowledge; thus, they rely on lecturing instead of promoting student discourse. Mentors should help protégés structure the classroom to encourage student communication and give students opportunities to help other class members.

Focusing on the "Big Picture"

Many new teachers follow the textbook section by section while missing the main objective of a particular chapter. Mentors should remind their protégés to make connections with the major ideas in a particular unit during instruction. By planning together, mentors can help protégés see important connections across lessons and units.

Using Creative Examples

Using obvious or repetitive examples during instruction can limit students' understanding of a particular topic. For instance, habitually discussing squares when referring to quadrilaterals in general may cause students to identify only squares as quadrilaterals. Here again, the mentor may wish to plan lessons ahead of time with the protégé while encouraging the new teacher to include a variety of examples and counterexamples.

Keeping Track of the Time Frame of Events

Mentors may record start and end times for all activities or segments of a lesson to help protégés manage instruction time effectively.

Varying Instruction

Protégés seem to use the lecture format for instruction most often, bypassing other effective approaches. The mentor-teacher may encourage the protégé to explore other strategies during a lesson.

Using Correct Mathematical Terminology

New teachers may sometimes say *squares* when they mean *rectangles* or use the term *equation* when they mean *expression*. If teachers frequently misuse terminology, students may also fall into the same habit. The mentor should clarify misused terminology and encourage the protégé to be cautious when speaking mathematically.

—Jason D. Johnson

Using Videotaping and Stimulated Recall to Reflect on Teaching

Using videotaping and stimulated recall with a coach or mentor is one way to prompt teachers to reflect on and improve their instructional practices. The following paragraphs describe one lesson in which stimulated recall helped a teacher find ways to enhance her mathematics instruction.

The Lesson

In my role as an instructional coach, I videotaped Tony while she conducted a lesson about equivalent fractions in her classroom. During the lesson, students were involved in a paper-folding activity in which they folded a piece of paper into thirds, then colored two parts of the three to represent 2/3. Students were then asked to fold the paper in half and talk about the resulting fraction. When asked, "What fractional part is shaded?" several students called out, "four-sixths." The teacher asked whether 2/3 is equal to 4/6. A few students said yes, a few said no, and one adamantly stated that the two fractions were not equal. The teacher discussed more examples with the students, using folded paper as a model. By the end of the class period, most students were beginning to develop a conceptual understanding of equivalent fractions.

Postlesson Conference

The postlesson conference with Tony revealed some important findings about her teaching. When watching the videotape several days later, Tony revealed, "I never expected to have to do so many examples. I thought they would get it on the first or second example." When I pointed out to her that this lesson perfectly illustrated the technique of allowing students to construct their own knowledge using manipulatives, she responded, "I hate to say it, but we are under so much pressure most of the time that we don't have time to teach this way. I can't believe how quickly the time went. Where do I get the time to do this kind of activity when the homework in the book just asks them to fill in the numerators and denominators?" As she continued to watch herself on videotape, Tony became frustrated; she explained, "I got confused myself. I was hooked into trying to get them to understand the algorithm. I was trying to fit the activity with the algorithm, and I myself had trouble with it. I was speaking during the lesson, but my brain was saying, 'Wait a minute; we are confusing issues here.' They did not have any practice finding missing numerators or denominators, the approach used in the text."

Tony and I discussed some of her instructional decisions, paying attention to an activity that she started at the beginning of the class. "What was the purpose of that activity?" I asked. "Looking at the lesson now, I realize it had no connection whatsoever," she replied. "I could have left it out altogether." When I asked how omitting the initial activity would have helped the lesson, Tony said, "I guess it would have left me more time to do the paper folding." I then asked, "Can you think of how you could have allowed the students to explore the paper-folding activity, then led them to practice filling in missing numerators or missing denominators with equivalent fractions?"

Benefits of Videotaping and Stimulated Recall with a Coach or Mentor

The videotaping session was an enlightening experience for Tony. Her uninhibited ability to reflect honestly about her instruction with a coach was beneficial in increasing her effectiveness as a mathematics teacher. Discussions about her lesson clarified actions she needed to take to improve her instruction. She knew she wanted to allow her students to construct their own knowledge of the concept of equivalent fractions but struggled to align her ideas with the textbook.

Often, teachers work in isolation from their colleagues and are judged by supervisors and administrators who visit and observe only sporadically. When conducted effectively with a coach or mentor, videotaped lessons and postlesson conferences allow teachers to examine their instruction objectively, discuss their instructional decisions, talk about their teaching, and draw conclusions. This teacher clearly took advantage of the videotaped lesson and the presence of the coach as a sounding board to investigate ways to improve her mathematics instruction!

—*Patricia A. Emmons*

Technology as a Communication Tool

Future teachers in the alternative teaching certification program in mathematics at Eastern Illinois University complete a teaching internship lasting one academic year. During the internship, candidates in the program are employed as full-time teachers by a cooperating school and are responsible for teaching a full load of classes and handling all related teaching duties. Program participants fulfill these responsibilities without the benefit of a cooperating or supervising teacher. Instead, candidates are assigned a mentor by the cooperating school. Owing to a lack of funds and staff and for other reasons, most participants have little to no contact with their assigned mentors. As I work with these beginning teachers, I try to fill the mentor role online using—

- online discussions;
- message boards; and
- reflective self-assessments.

The use of online discussions and message boards allows program participants to read advice and post comments at times convenient for them. These tools also enable a timely response system with better flow and organization than e-mail, provide a running record of the discussions that have taken place among participants, and help the novice teachers track their progress. The data gathered online give evidence of growth and change during the participants' teaching experiences and become part of a database of information to draw on to help future candidates in the program.

I use reflective self-assessments as conversation starters that encourage candidates to think about issues related to teaching. These self-assessments are posted online so that candidates have time to organize their thoughts before posting responses. Below are sample reflections I have used to initiate discussions.

- How are you handling day-to-day issues related to teaching mathematics? Do you feel you have time for yourself?

- Describe how you came to your current classroom-management plan. How is the plan different from what you initially imagined?

- How are you handling homework issues? Do you assign homework? Do you grade it? If so, how? If not, why not?

- Describe a lesson that you felt was particularly successful. What made the lesson work? Did anything unexpected happen during the lesson? How much planning time did the lesson require?

Many program participants indicate that they appreciate this type of mentoring because it does not make them feel pressured or intimidated. They feel comfortable posting and responding to questions and having discussions online rather than face-to-face. More than one new teacher has said that the online mentoring system served as a consistent lifeline for help and advice that they could not get from their on-site mentors.

—*Marshall Lassak*

Mentoring through Cognitive Coaching

As a mentor, I strive to build trust while using a cognitive coaching model to support the many challenges of the inexperienced mathematics intern. Cognitive coaching suggests that supporting novices in reflective, self-directed learning is a sure path to improved performance (Tabor 1998). The cognitive coaching model

requires the mentor to observe, provide opportunities for reflection, and offer insights when requested. Time, respect, and a nonjudgmental attitude are required to build trust in the mentoring relationship. Ultimately, the cognitive coaching model increases independence for mentees and establishes the habit of reflective teaching. The following sections describe a case study in cognitive coaching in which the mentee learned about controlling classroom behavior, addressing varying ability levels, and engaging students with the mathematics lessons.

Controlling Behavior

I met Ms. G. in her eighth-grade mathematics classroom. She had already been teaching mathematics for a month. The school district had recommended that Ms. G. spend the first eight weeks reviewing skills that the students had not mastered in previous years of mathematics instruction. Her goal for the year was to teach first-year algebra, but she had not even started to introduce new topics.

Ms. G. stood in front of thirty-eight talkative students who paid little attention to her continued efforts to reteach concepts involving fractions. I spent an hour observing her and later asked her to reflect on her teaching. The first concern she had was classroom management. Reflecting on her management style, Ms. G. decided that it was not working because she was not consistent. She needed to institute a behavior plan with rewards and consequences. Ms. G. developed the plan, asked for my suggestions, and decided to implement the new procedures the following week.

Addressing Varying Ability Levels

Behavior was just one challenge that Ms. G. faced, however. She also struggled with the varying ability levels of the students in her class. She asked, "How can I teach more abstract ideas in algebra when students do not understand basic concepts?" I listened as she described the kinds of lessons she had taught. When she asked me for suggestions, I recommended that she try to provide support for basic concepts, a technique known as scaffolding, while teaching new material. I also advised her to put her students into cooperative groups for peer support. I asked her to think about creating groups that would provide peer support for students struggling in mathematics, those learning English, and those who caused behavior concerns. Ms. G. knew her students and carefully thought about group placement. In her final reflection, she wrote, "Having students working in groups was very beneficial for them and for me. Although some of the students were talking about something else, most of them were really helping each other with the mathematics and explaining it to each other in words that made more sense to them."

Engaging Students with Mathematics Lessons

A month later, I returned to the classroom for another observation session. The behavior plan and cooperative group work had improved classroom management, but Ms. G. commented that her students were not as engaged with the lessons as she would have liked. She said, "One of the hardest parts of being a teacher is dealing every day with the frustration of having students who refuse to use the smallest effort and with the guilt that you as a teacher do not do enough to motivate those students." I listened to her as she struggled to identify why her students were not motivated. Ms. G. noted, "The hardest part for students is thinking." She said that her tasks did not require higher-level thinking, and that her students were not challenged and seemed bored. I asked Ms. G. whether she could modify the tasks, explaining that problem solving should be "a primary goal of all mathematics instruction and an integral part of all mathematical activity" (NCTM 1989, p. 23). Thinking about her own experiences in professional development courses, Ms. G. decided to give her students tasks that would be more challenging and would require higher-level thinking. We spent some time discussing tasks and mathematics pedagogy, and Ms. G. was eager to make changes in her classroom.

Concluding the Coaching Experience

Over the course of the school year, I observed Ms. G. several times, listened to her reflections, and offered suggestions when asked. On my final visit, Ms. G. commented, "Everything is about expectations." She explained that she believed teachers should have high expectations for behavior and academic achievement. Ms. G. continued, "Seeing my students multiplying or factoring polynomials makes me see that everything is about expectations. The higher they are, the better. The majority of my students did not have any problem understanding algebra, which I thought they would never be able to understand."

Ms. G. concluded the school year by asking her students to write a letter to next year's students. A student named Mayra wrote, "Algebra is a really fun subject to learn about. You will get a chance to play with different kinds of stuff for different kinds of algebra problems." Another student, Richard, wrote, "While learning new material you will have fun doing it." Ms. G. concluded her final reflection by saying, "Having students tell me, 'I feel so smart, Ms. G.' because they are solving problems that are not easy gives me the incentive to go on."

—Deandrea L. Murrey

BIBLIOGRAPHY

National Council of Teachers of Mathematics (NCTM). *Curriculum and Evaluation Standards for School Mathematics*. Reston, Va.: NCTM, 1989.

Tabor, Marilyn, presenter. Coaching Conference with a Focus on Literacy: Professional Interaction Founded upon the Principles of Cognitive Coaching: An Introduction for University Associates and Student Teachers at Professional Development Schools. Master Teacher training session, University of California, Irvine, December 1998.

Talking about Teaching: A Strategy for Engaging Teachers in Conversations about Their Practice

One challenge faced by mentors is how to provide instructional support that enhances performance without judging or criticizing beginning teachers. Fostering a safe environment for conversations to take place is essential. At the same time, the mentor's use of nonthreatening language is important to sustain regular discussions about instruction. Consider, for example, the exchange that took place between Olivia, a preservice teacher completing a full-year internship in a middle school classroom, and her university supervisor, Nora. Nora was deliberate in her choice of words as she helped Olivia reflect on her teaching.

The Lesson

Olivia had just taught a lesson on finding the slope of a line represented in each of three forms: table, graph, and equation. As she launched the lesson, she drew the sketch shown in figure 4.1 on the board, reminding students that slope is the ratio of the rise to the run. As she drew the sketch, she correctly indicated that rise was the vertical change and run was the horizontal change. This distinction, however, was not clear in the completed drawing. As students worked in small groups to explore an assigned task, they exhibited some confusion about how to determine the slope from a table of values or a set of ordered pairs. Olivia patiently asked students questions about their work: What is the rise? What is the run? What is *m*? What is *b*? What is the equation? Still, some students did not understand what values to use in forming the ratio.

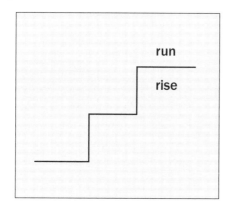

Fig. 4.1. Olivia's drawing

The Postlesson Conference

During the postlesson conference, Nora asked Olivia for her thoughts about the success of the lesson. Olivia noted that the students seemed to confuse rise and run, but she was not sure why. Nora pointed to a sketch she had made in her notes similar to the one that appears as figure 4.1 and told Olivia that she had *noticed* the diagram on the board at the beginning of class but *wondered* whether the labeling of the diagram might have been a source of the students' confusion. Nora's comments led to a conversation about how students might have misinterpreted the diagram as "run over rise." Nora and Olivia discussed other representations that might have helped students make sense of the concept of slope and the strategies for finding it.

Nora's deliberate use of *noticed* and *wondered* serves as an example of a nonthreatening way to talk about instruction. A mentor who *notices* something in a classroom draws on evidence from the observation session, for example, student work or field notes. The the mentor makes only factual statements, avoiding evaluation or personal preferences. Sometimes, the mentor may notice instances of good instructional practice, which should be reinforced with the beginning teacher to ensure that such practices are repeated. During the discussion with Olivia, Nora also noticed that Olivia had asked questions that made students think, as evidenced by their responses. For example, Olivia had asked, "If I give you a graph and you want to find the slope, what would you look for?" One student responded, "I would see if the slope was positive or negative by seeing which way it slanted," and another said, "I would make steps between the points and see how much it went up and over." Because questioning was something that Olivia had been working on, Nora wanted to make sure that she reinforced this aspect of the lesson.

In some instances, mentors may notice a practice that needs to be refined. The mentor should identify the practice and support his or her observations with evidence before engaging the beginning teacher in a discussion about how to improve his or her instruction. The use of the word *wonder* serves as a nonthreatening approach to launching this discussion. In this example, Nora used the word to engage Olivia in discourse that focused on reflection and problem solving related to representations of slope, an area of Olivia's practice that needed refinement. Engaging in such discussions may be the most difficult part of the mentoring process, but it is essential if growth is to occur.

The Noticing and Wondering Model

Mentors should structure a rigorous examination and analysis of practice by considering the beginning teacher's decisions and discussing both positive and negative results of instructional practice. The *noticing* and *wondering* model, shown in figure 4.2, is one way to promote such examinations of practice.

—*Margaret Smith*

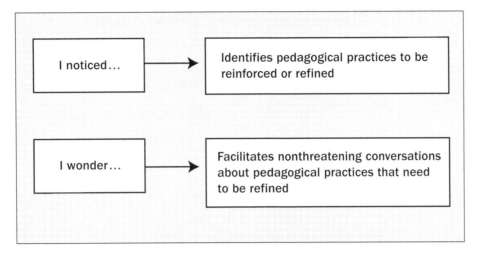

Fig. 4.2. Model for noticing and wondering

A Guide for Reflecting on Mathematics Lessons with Beginning Teachers

The beginning teacher you are mentoring has invited you to observe a mathematics lesson. Great! Now comes the hard part. How do you begin to discuss the lesson and stay focused on mathematics? Too often, issues of behavior or timing get in the way of discussing the mathematics in the lesson. One way to approach the discussion is to use the mathematics lesson reflection guide (see fig. 4.3), which incorporates the five Process Standards from *Principles and Standards for School Mathematics* (NCTM 2000). Asking about evidence of success concentrates the beginning teacher's reflections on tangible proof that students understand the mathematics being taught, and thinking about missed opportunities is a positive, less threatening approach to discussing needed improvements (Van Zoest 2004). Using the guide in mentoring situations enables richer discussions between the mentor and the protégé that stay targeted on important mathematics.

—*Nancy O'Rode and Hillary Hertzog*

REFERENCES

National Council of Teachers of Mathematics (NCTM). *Principles and Standards for School Mathematics.* Reston, Va.: NCTM, 2000.

Van Zoest, Laura R. "Preparing for the Future: An Early Field Experience That Focuses on Students' Thinking." In *The Work of Mathematics Teacher Educators: Exchanging Ideas for Effective Practice,* AMTE Monograph 1, edited by Tad Watanabe and Denisse R. Thompson, pp. 119–34. San Diego, Calif.: Association of Mathematics Teacher Educators, 2004.

Mathematics Lesson Reflection Guide

Provide specific examples from the mathematics lesson for each of the boxes.

Process Standard	Evidence of Success	Missed Opportunities
Problem Solving Students actively engaged in problem solving.		
Reasoning and Proof Students made sense of the mathematics by explaining their reasoning and justifying their thinking.		
Communication Students expressed their mathematical thinking to others.		
Connections Students connected or applied mathematical ideas in appropriate contexts.		
Representation Students used multiple representations to model mathematical ideas.		

Fig. 4.3. Reflection guide based on NCTM Process Standards

Promoting Equity in the Mathematics Classroom

How do teachers promote equity in the mathematics classroom?

Why is reflecting on the issue of equity in the classroom important for teachers?

What are some characteristics of equitable mathematics classrooms?

In an effort to respond to these questions and help teachers think about equity issues in their practice, we developed an initial set of reflective questions. We further refined the questions through discussions with a group of teacher-leaders who were responsible for mentoring beginning teachers in mathematics.

We first identified four purposes for promoting equitable classroom practices, as follows:

- To provide access to mathematics for every student

- To develop the intellectual capacity of each student

- To encourage students to share their thinking and develop confidence as learners of mathematics

- To establish a classroom environment that is inclusive and respectful

The reflective questions in figure 4.4 (on the following page) are organized into groups corresponding to these four purposes. Teachers should use these questions as they plan lessons, reflect on their practice, or think deeply about the classroom environment. In our project, some teachers chose to focus on one or two questions at a time, using them to conduct informal research in their classrooms. Other project participants used the questions to promote discussion about important elements to consider in addressing equity in the mathematics classroom. The teacher-leaders found the questions to be useful in reflecting on their own practice and in promoting discussion with beginning teachers, but they agreed that the questions should not be used as a tool to evaluate protégés.

—*Nancy Terman, Nancy O'Rode, and Maria Guzman*

Equity in the Mathematics Classroom: Reflecting on Your Own Classroom

The following questions relate to equity in the mathematics classroom and have multiple uses. Use them in planning or reflecting on lessons and thinking about your practice or classroom environment. We suggest that you focus on only a few questions at a time, and keep in mind that this list is not comprehensive; you are encouraged to add your own questions.

Providing Access to Mathematics

Prior Knowledge

How does my lesson help students retrieve their prior knowledge about the topic?

Target Vocabulary

Are my students given opportunities to understand and use target vocabulary in ways that make sense to them?

Hands-on Activities

Does my lesson give all students the opportunity to work with manipulatives and other concrete learning tools?

Language

How does my lesson provide access for students at different reading and language levels and enable them to be fully engaged?

Learning Styles

Does my lesson consider the multiple and varied learning styles of students?

Developing Intellectual Capacity

Critical Thinking

How does my lesson promote higher-order problem-solving skills and intellectual rigor?

Mathematical Ideas

Do my students learn important and challenging mathematics through my teaching?

Mathematical Reasoning

Does my lesson allow students to question, make conjectures, and justify their mathematical thinking?

Expectations

How do I communicate and maintain high expectations for every student? Are all students challenged intellectually?

Encouraging Student Thinking and Discussion

Lesson Climate

How does my lesson help students develop confidence as mathematical learners and encourage individual thinking and risk taking?

Student Engagement

How does the lesson engage students? Are all students engaged? What are the opportunities in my lesson for students to work collaboratively, in small groups or pairs?

Student Participation

What efforts do I make to include all students in class discussion, especially those who are quiet or passive? Do I follow established rules for participation, such as enforcing a wait time to ensure that no one student dominates class time and teacher attention?

Developing a Culture of Respect and Inclusion

Meaningful Context

Does my lesson provide a meaningful context; that is, is it connected with students' background, culture, or experiences? Do students see what they are doing as important for their lives?

Culture of Inclusion

How is an overall culture of inclusion and expectation encouraged in my classroom? Do I explicitly espouse and practice inclusion? How does the physical environment of my classroom promote inclusion?

Climate of Respect

Do I promote and maintain a climate of respect for students' ideas, questions, and contributions in my classroom? How do I encourage students to bring their own ideas, thinking, and experiences to their learning?

Language and Behavior

Does my language and behavior clearly demonstrate sensitivity to issues of gender, race or ethnicity, special needs, English-language learning, culture, and socioeconomic status?

Challenging Bias and Stereotypes

Do I take opportunities to recognize and challenge stereotypes and biases that become evident among students? Do I analyze my interactions with students to check for biased language and stereotyping?

Fig. 4.4. Equity reflection questions

Section 5: Collaboration as Mentoring

MENTORING is inherently about collaboration. The mentor collaborates with the protégé with the ultimate goal of making a positive impact on student learning through the professional growth of both members of the team. In the teaching and learning of mathematics, the purpose of collaboration is to come together to share ideas and to construct new learning. Collaborating in teaching involves creating a relationship in which participants ask why certain strategies are used, why certain outcomes occur, and what might happen if other approaches are introduced.

On the one hand, collaboration in a mentoring relationship is not significantly different from collaboration between experienced teachers of mathematics. In both situations, the parties seek to create shared understanding as a means to increase knowledge and learn together. On the other hand, the purpose of collaboration in mentoring is to support the growth of the beginning teacher. In this way, the nature of the mentoring relationship influences the framework for the collaborative process. Although the possible structures for working with others are endless, the following articles offer various suggestions about how mentors might collaborate with one another or how mentors and protégés can work together in a professional partnership.

66 The mentoring relationship... is one of interdependence, trust, and a quest for mutual growth. **99**

—*Sarah Wallus Hancock and Cheryl Ann Lubinski*

39

Mentors as Learners: A Practice-Based Approach to Mentoring

Learning to teach is both an individual and a social activity, and both halves of the equation are essential to professional growth. Mentoring provides the opportunity to engage in both kinds of learning activities on two levels: First, mentors and protégés question and reflect on their own practice individually, and second, through their sharing, contribute to the enrichment of each other's practice. Likewise, when a group of mentor-protégé pairs meets regularly to discuss their experiences, the isolated work of the pair becomes a social activity and the seeds of a learning community are planted.

The activities discussed in the following sections may be useful in stimulating learning among both mentors and protégés. Mentors should note that modeling learning through their practice increases their effectiveness with beginning teachers.

Learning through Shared Practice

Mentors and protégés alike can learn more about their teaching practices from jointly investigating the work of teaching. Shared practice offers rich opportunities for reflection and learning because it provides common experiences for the mentor-protégé pair to discuss and analyze. Mentors and protégés may become involved in shared practice through coplanning, coteaching, collaborative examination of student work, and joint observation of other classrooms. An example of coplanning a geometry lesson illustrates the benefits of this approach for both the mentor and protégé and highlights the mutual contribution of ideas that is inherent in the meaning of *collaboration*.

A mentor-protégé pair collaborated to plan a high school geometry lesson on triangle centers. They approached the planning task by asking themselves a series of questions: What if we introduce the concept in this way? How might students respond if we ask this question? In this collaboration, the mentor-teacher gave insights into common student struggles and misconceptions while the protégé contributed his expertise in deductive explanation and the computer program The Geometer's Sketchpad. In this situation, both the mentor-teacher and protégé were learners, with

the mentor sharing her experiences but also participating in reflective conversations about her own practice.

Learning through Community Sharing

Mentor-protégée pairs can benefit from making their work public within a professional community. In one program, groups of three to six mentor-protégé pairs met weekly or bimonthly. The facilitator's role was to stimulate learning within and between the pairs by suggesting activities and carving out time for conversations in the hectic schedules of teachers. The group meetings generally focused on the collective investigation of teaching artifacts; for example, group members watched videotaped lessons, analyzed student work, and solved mathematical tasks to be used in the classroom. The following example highlights the possibilities for community development in this setting.

During a cluster meeting, a mentor-protégé pair shared a mathematical question that a student had posed in a calculus class. In the course of a lesson, one student noticed that integrating is reversing differentiation. This comment prompted another student to ask, "Does every mathematical process have an inverse process?" Recounting this mathematical situation to the group launched an exploration of whether reversibility is always, sometimes, or never true. The ensuing discussion touched on a range of mathematical activities, from thinking about if-then statements in a geometry course to solving quadratic equations in algebra. Toward the end of this discussion, teachers suggested possible responses to the student's question. Through this social activity, all participants learned new ways to think about reversibility.

Note that the mentor-teachers participated in these meetings as learners. In this role, they gained new knowledge about their own teaching practices and modeled learning from practice to enhance the mentoring process for their protégés.

Individual and Group Learning Combined

Positioning mentors as learners within a community distributes responsibility in the mentoring relationship more evenly and enables all participants to grow.

Individual learning contributes to group learning, creating a whole that is much greater than the sum of its parts.

—Ginger Rhodes and Patricia Wilson

A Tale of Two Mentors

Audrey, a relative newcomer to the profession in her third year of teaching, is responsible for a class of third graders, and Cathy is a teacher educator. The two met when Audrey participated in a series of professional development sessions led by Cathy to support teachers in using their students' mathematical understanding to drive instruction. Since that time, the two have forged a "co-mentoring" relationship through which they collaboratively teach in Audrey's classroom, discuss students' thinking as a means of planning instruction, and work together to ensure the development of students' mathematical understanding.

As the primary classroom teacher, Audrey conducts ongoing assessment of her students' thinking and shares this information with Cathy. In return, Cathy keeps Audrey current with research relevant to mathematics teaching and assists in developing Audrey's pedagogical content knowledge. The mentoring relationship for these two professionals is based on interdependence, trust, and a quest for mutual growth.

Comments from Audrey and Cathy

Audrey. As a mentee, I must feel comfortable asking questions on such topics as the use of proper mathematical terminology and mathematically correct representation, sources of relevant research, and strategies for analyzing students' thinking to make instructional decisions. I must know that my questions will be answered with expertise, but I will benefit even further if that expertise is tempered with an approach that says, "Let us think this through together." This process allows me to move through the steps of inquiry with my mentor instead of receiving pat answers to my questions or simply being advised of the "correct" ways to address issues that undoubtedly will arise in the mathematics classroom.

Cathy. Before graduate school, I taught elementary school for fifteen years. At that time, I had few insights about how to develop students' reasoning and very little knowledge of how to plan for instruction using students' thinking or how to teach for in-depth understanding. Because I did not return to the elementary classroom after graduate school, my opportunities to acquire knowledge about elementary students' learning trajectories and how those trajectories could be developed were limited to visiting classrooms over several days to observe mathematics lessons. Audrey helps me focus on her class as a whole and on her students' knowledge, as well as how that knowledge can be developed, especially individually. As part of this process, I have learned better questioning techniques. What's more, I am enjoying teaching at the elementary level once again!

Benefits of the Co-Mentoring Relationship

School cultures that assume students learn best through a process of inquiry and discovery often extend this assumption to the learning of teachers. That is, teachers who seek to extend and develop their knowledge through an informal mentoring relationship are supported and encouraged by those who realize that the fruits of the process will extend beyond depth of understanding for teachers to depth of understanding for students.

The authors' co-mentoring relationship is successful because we have an interest in learning and developing our own expertise. As mentees, we play the role of learners by default, but as mentors, we send a powerful message that we are also continually learning. We try to be straightforward about our expertise but honest about what we do not know. As our co-mentoring relationship develops, an important method for ensuring professional growth for both of us is the process of co-reflection and debriefing about classroom practices, student thinking, research, pedagogy, mathematics content, and the interconnectedness of all these aspects of teaching. In this way, the tale of two mentors continues.

—Sarah Wallus Hancock and Cheryl Ann Lubinski

Collegial Mentoring

As a teacher-leader in elementary mathematics at both the school and district levels, I wanted to learn whether mentoring in a specific subject area would benefit a colleague and support a process of change. Sophie, a teacher and self-identified sufferer from "math anxiety," wanted to change her beliefs about mathematics. She agreed to submit to observations and interviews and to keep a journal in which she would record her feelings about mathematics. Her goal was to gain greater understanding of her anxiety, which she believed would help her "let go" of it.

Sophie's Anxiety

During one observation session, Sophie had planned the lesson independently and allowed me to observe to give feedback on her approach to teaching. Being observed was also a way to encourage Sophie to teach mathematics more often. Although the lesson I observed was satisfactory, I could tell from the flush on her neck and her avoidance of eye contact that my presence made Sophie uncomfortable. She wrote in her journal, "Being observed yesterday in math was hard. I did better without being observed on a day when I was very tired. I didn't like being asked unpredictable questions in front of [the observer], and I had a fear of being judged…."

Sophie still felt unsure about her ability to teach mathematics effectively, and as her mentor, I needed to foster a greater sense of security for her in our relationship. Throughout the partnership, Sophie continued to experience mixed emotions about mathematics and the mentoring relationship in general. She struggled with, and continued to be frightened by, mathematical concepts. Not surprisingly, those feelings hindered her ability to teach the subject and negatively affected her self-image as a teacher.

Family Math Night was another source of anxiety for Sophie, or as she wrote, "not my idea of a good time." After agreeing to coplan the activities, Sophie informed me that one activity in particular caused anxiety for her. When faced with a mathematical concept she could not grasp, Sophie believed that she could not help her students and, thus, became anxious and disappointed in herself. After I encouraged her to figure out the solution

to the activity with scaffolding, we agreed to alter the activity to include only the part she felt confident about, and she was relieved. At the end of the evening, she told me, "I survived. The best thing about the evening was watching parents and kids so involved, working together. I missed that collaboration as a child."

Growing Enthusiasm

One Sunday, Sophie called me at home to set up an interview and observation session. She was enthused about a unit she was planning that involved the Olympic Games, and she described it to me eagerly: "That's the stuff I get excited about!" This real-life event helped motivate her to create some interesting lessons in mathematics to which she felt connected. Sophie said, "I like teaching mathematics a lot better using the Olympics, and where there is meaning and worth for me, there is meaning and worth for the kids."

Over time and with encouragement, support, and feedback, Sophie began to gain confidence in herself and her abilities as a mathematics teacher. She said, "Enthusiastic and encouraging math teachers made all the difference to me.… In the last two years with [my mentor] and [another colleague], I have experienced a lot more support and a team approach…." In her journal, Sophie wrote, "…I appreciate the support and [my mentor's] positive feedback and presence, her being close and being there."

Outcome of the Partnership

As a result of the positive feedback Sophie received in the mentoring partnership, she came to realize that she was a much stronger mathematics teacher than she had originally thought. Mentoring helped her realize that she could offer constructive lessons to her students in mathematics. Sophie firmly believed that all her students should have confidence in their abilities, whether they were strong in art, language, mathematics, or another area; through mentoring, she was also able to transfer this belief to herself. For Sophie to maintain her positive attitude toward mathematics, she will need to continue reflecting on, and refining her beliefs about, the subject and will be helped by collegial support, feedback, and encouragement.

—Jennifer Wyatt

A Sisterhood of Mathematics Teachers

As a group that includes both veteran educators with more than twenty-five years of public school experience and novices in their first years of teaching, we know firsthand the value of mentoring. In our school, we have formed a sisterhood of teachers, primarily in mathematics, to serve as mentors to one another. When we first met, some of us had teaching experience, some were new to the profession, and others were simply new to teaching middle school mathematics. Some of us have worked together for many years, whereas others have been on campus for only a year or two. The mentor-mentee relationship we have with one another is a two-way street that has sustained all of us both professionally and emotionally over the years. We believe that this relationship is the primary reason most members of our sisterhood have remained active in education. We are also confident that others can establish or encourage a "sisterhood" mentality in their unique settings.

Overcoming Isolation and Cultivating Leadership

Many newcomers to the teaching profession have no idea how important mentorship is to their professional success, not to mention their mental and physical health! Unfortunately, some "stuck in a rut" veteran teachers are also unaware of the professional benefits of the mentor-mentee relationship. Those of us who started teaching before formal mentoring was conceived know the anxiety and isolation that newcomers to the profession face.

Cultivating educational leaders, the primary goal of any mentoring program, has been one of the most fulfilling aspects of our sisterhood. Our commitment to helping one another develop individual strengths without giving negative or critical feedback has served us well. The unbreakable bond we have formed not only supports the teachers in the group but also serves the needs of students and educational stakeholders and creates a synergistic environment. Early on, we relied on one member of the group as the leader, but now, many of us have developed areas of particular leadership expertise. Our campus colleagues also benefited from our mentoring activities when, as a department, we lobbied for and received a common departmental planning period, an innovation that is now a standard for all core subjects. The campaign for the common planning period was foundational to the evolution of our sisterhood because it gave us opportunities to learn leadership skills that we have since successfully transferred to other settings.

Exhibiting the Qualities of Good Mentors

We acknowledge that our sisterhood has not been perfect in all aspects. Unfortunately, we have learned that not all veteran educators make good mentors. At the same time, one of our members who teaches outside the mathematics department believes that her informal mentor epitomizes the best qualities of those who play this role. Unlike some of the unskilled mentors we have encountered, this one does not tell her mentee how to teach, ostracize her for running her classroom differently, or force her to adopt the mentor's methods. Rather, this mentor allows her mentee to make mistakes and only then asks reflective questions that focus on a few significant aspects of the mentee's teaching.

Achieving Independence and Growth

A true mentor is a "cognitive coach" who facilitates reflective thinking and arms the mentee with the skills to reach independence. Our influence on one another's educational thought processes occurs along a spectrum from informal hallway conversations to formal mentor meetings. Openness, trust, and honesty are essential ingredients in this relationship and unique characteristics of our sisterhood. Helping one's colleagues, whether veterans or novices, reflect on their teaching pays much greater dividends than giving them all the answers.

We in this sisterhood know the benefits of mentorship. We continue to support one another by attending members' workshops, celebrating promotions and the completion of advanced degrees, and encouraging one another to contribute to our profession in unique ways. We appreciate those who have mentored us through their words and deeds and look forward to "paying it forward" to the next generation of educators.

—Shirley M. Matteson, Nancy Box, Cate Hartzell, and
Roxanne Howell

Collegial Mentoring through Lesson Study

Historically, the practice of teaching has been an isolated field. Teachers go into the classroom, close the door, and interact with students. Interaction and learning with other adults is fragmented and rarely applied systematically to classroom lessons. Little time is available in the structure of the educational system to support site- or district-level sharing about issues related to teaching.

Benefits of Joint Lesson Study

By engaging in joint lesson study, teachers can distribute the demands of their jobs more widely and collectively construct solutions that have an impact on multiple classrooms rather than a single classroom. Each teacher contributes unique perspectives, experiences, and background knowledge to the process of lesson study. In turn, the practice of sharing various beliefs and experiences prompts teachers to think in different ways. For example, during a planning meeting for fifth grade, one teacher mentioned the desire to provide hands-on opportunities for students to understand fractions. Another teacher wanted the students to be able to visualize fraction amounts. The combination of the two perspectives enabled the creation of a lesson that moved students from concrete to abstract concepts and provided scaffolding that met the needs of both students and teachers.

Without the lesson-study process, the sharing of ideas is limited; however, as evident in this example, teachers benefited from the exchange of new ideas and passed those benefits on to their students. Through the lesson-study process, individual and unique knowledge banks are combined and leveraged to construct better teaching practices. Teachers develop a common understanding of learning problems and common reasoning to support teaching solutions. If a particular lesson shows weakness, the group uses evidence gathered during instruction to carefully refine it. By applying multiple perspectives to the problem, the group enhances the lesson to achieve maximal learning.

Collegial Mentoring for Novices and Veterans

The need for collegial mentoring applies to both novice and veteran teachers. Both groups improve their communication skills, teaching practices, and subject-matter understanding as they work with other professionals. Veteran teachers are energized by the process of collegial problem solving and the appreciation of the group for their knowledge and experience. Hearing affirmation that their approaches are sound builds confidence and invigorates teachers to continue improving their teaching practices.

New teachers are also motivated by the lesson-study process, and the pace of their learning is accelerated. By listening to, and participating in, focused curricular conversations, new teachers gain insight into proved practices, ideas, and methods. They have access to a pool of experienced teachers from whom they may seek advice in lesson planning. In this way, new teachers begin to develop tools and strategies of their own to create effective lessons. Having the opportunity to plan and reflect with experienced teachers allows beginning teachers to build confidence and gain knowledge in a safe environment while veteran teachers elaborate on and expand their teaching practices.

School-Specific Results

The lesson-study process allows teachers in my school to collaborate while maintaining the focus on student improvement. Natural mentoring relationships evolve as group members work to understand the complexities of teaching and learning. The group understands that the purpose of lesson study is to learn and improve. All teachers, beginners and veterans alike, benefit from examining their own teaching, and the combined ideas of all group members enable us to build better lessons. Through collaboration, we teachers have emerged from isolation and have the opportunity to share with, and learn from, others in our profession.

—*Susan Mulligan-Hirsch*

Team Mentoring: Cooperation and Success without Consternation

When a team of teacher-educators who specialize in different educational areas, such as a special education teacher and a mathematics teacher, mentor a group of in-service teachers, the success of their collaboration depends on moving toward mutual goals for the benefit of the mentees. To be successful, such a team-mentoring model should exhibit the following characteristics:

- Each mentor's specialization should be valued.

- The mentors should engage in the free exchange of ideas.

- Disagreements should be viewed as an opportunity to expand one's thinking.

- The mentors should seek feedback on the direction of the mentorship to preserve aspects of the process that are most effective in supporting change.

Integration of each mentor's skills and opinions is essential when team mentoring. In addition, team mentoring, when structured carefully, offers an opportunity for mentees to witness and experience collaboration among professionals.

The process of team mentoring begins with the mentors discussing what each believes are the needs of the mentees and reaching a consensus about the goals for the mentoring process. For example, one team suggested the following goals for a group of mentees:

- Hold discussions that give teachers the opportunity to analyze students' work.

- Develop strategies for inquiry-based learning or discovery learning of mathematics.

- Develop strategies for differentiated instruction in mathematics.

- Fortify the mentees' mathematical content knowledge and strategies for using pedagogical content tools.

After agreeing on the goals, the mentor team provides support according to each member's area of specialization. None of the specialties is considered intrinsically more important than the others, but each may take precedence at one time or another during the mentoring process. The advantage of this model is that it offers diversified support that can be adapted easily to meet the changing needs of the mentees.

—Danté A. Tawfeeq

Section 6:
Ideas for Mentoring Programs

WHETHER your school or district has an existing program or is seeking to create a program, ideas for structuring mentoring arrangements abound. Some schools or districts have formal mentoring programs. In other situations, individual teachers take on the job of mentoring unofficially. The ways to approach mentoring are as varied as the teaching and learning styles of ourselves and our mentees.

At its core, mentoring is focused on the continual development of professionals. More specifically, in mentoring relationships for mathematics teachers, both the mentors and protégés continue to develop their content knowledge of mathematics, increase their pedagogical knowledge, and seek ways to blend the two to enhance students' learning. Both parties benefit from the mentor-protégé relationship, and for this reason, mentoring programs can be designed to include components that target the professional development of mentors as well as the growth of beginning teachers.

Regardless of the vision and goals that your school or district may have for a mentoring program, the collection of articles here offers a plethora of ideas and suggestions, with topics ranging from criteria for creating mentoring programs to the professional development needs of mentors.

66 Mentors come in all ages and sizes, and with backgrounds as colorful and diverse as the signs in Times Square. 99

—*Heather A. Martindill*

Criteria for Teacher Induction, Mentoring, and Professional Development Programs

Across the nation, the emphasis on mentoring has put pressure on districts and school systems to institute mentoring programs. In turn, this pressure has created, in some instances, a "starting from scratch" or "beg and borrow" mentality that has resulted in fragmented professional development. To alleviate this problem, researchers, administrators, and field-based practitioners must analyze the criteria for success in this venue of professional development. Measures of success in mentoring programs must align with teacher retention rates, assessments of teacher competence, and ultimately, student achievement. The criteria listed below incorporate research-based elements that can be analyzed and used to develop and refine induction and mentoring programs to provide a comprehensive support structure for teachers. These criteria can also be used to support partnership models with universities and within individual district or school settings. Note that a general category is listed for each criterion, followed by a question to be answered to assist in verifying its fulfillment. The criteria and questions are adapted from Kortman (2005).

Criteria for Induction, Mentoring, and Professional Development Programs

- Infrastructure for support: How is the support structure of the mentoring or induction program defined?

- Communication of defined roles: What are the outlined responsibilities of all stakeholders in the program?

- Research-based, ongoing program development: What research supports the elements that are being developed or refined in the program?

- Alignment of program components with teacher population: What are the developmental needs of the teachers being served in the program, and how does the program address those needs?

- Program evaluation and feedback: What elements of the program are measured, and what tools are used for assessment?

- District-, school-, and classroom-embedded implementation: In what ways is the mentoring program incorporated into the school system and the individual teacher's job?

- School culture of support: What scenarios illustrate support of the program at the school level?

- Equity for all teachers: Do all teachers have access to support, feedback, and ongoing growth opportunities?

- Professional development aligned with teaching standards: What professional teaching standards are addressed through the mentoring or professional development programs offered?

- Alignment with certification requirements: Are professional development programs designed to help teachers achieve highly qualified status as professionals?

- Collaboration in a learning community: What criteria are necessary for effective collaborations in a learning community?

- Ongoing teacher leadership development: How are teacher-leaders and instructional coaches supported in their ongoing professional development?

- Beginning teacher support system: What elements create a comprehensive system of support to strengthen the work of beginning teachers and enable growth in teaching skills?

- Differentiated support for career growth after the second year: What professional development opportunities related to teaching standards are offered to ensure sustained growth and reflective practice over time?

- Mentor-teacher training system: What skills are taught to enable mentors to work effectively with teachers at all levels of development?

- Classroom visitation coaching: What are the goals and timelines for support provided through instructional coaches?

- Documentation of teaching competence: What data are gathered to verify teachers' growth through induction and mentoring support?

- Evidence of impact on students' achievement: How are student achievement data analyzed to correlate students' success with teachers' competence?

Outcomes of Professional Development Programs

The cycle of growth generated through a comprehensive professional development system positively affects both collegial and teacher-to-student relationships. Administrators repeatedly report that the culture of their school communities changes for the better when high-quality mentoring programs are established. These changes are attributed both to the support system among colleagues and to the individual growth of mentors and mentees in their practice. With such changes, the hope of mentoring support is realized. The level of confidence and competence grows among teachers, along with their commitment to the profession, to the development of relationships with educational colleagues reaching for joint goals, and to the success of students. Not only do teachers stay in the field of education, but they contribute in meaningful ways to their educational communities, so much so that teacher-leaders are often cultivated from within the system. Most important, students benefit from teachers who continually strive to enhance the quality of their practice.

—Sharon A. Kortman

REFERENCE

Kortman, Sharon A. "Building Competence in Teaching through a Systemic Induction, Mentoring and Professional Development Model: From Support to Accountability." In *Proceedings of the Teacher Quality Enhancement Grant Programs Conference,* edited by J. Middleton, C. Vallejo, and J. Kim. Tempe: Arizona State University, 2005.

Forming a Cadre of Mathematics Mentors

One of the goals of our school district was to give teachers the necessary professional development support to foster the teaching of mathematics for understanding. The strategy to make this profound change districtwide was to form a core group of mentor-teachers.

Project Structure

Mentors participated in two phases of the project. The first phase took place during the spring, and the second phase, during a summer session. In the spring, fifteen mentors took a graduate-level course in mathematics methods for credit, focusing on teaching in the middle grades. The course was offered at one of the schools in the district and met for three hours each week over fourteen weeks. The coursework gave special attention to topics that are important in the middle grades, such as number sense and place value; meanings of operations and operations with integers; fractions, rational numbers, and decimals; transitional concepts for algebra; ratio, proportion, and similarity; and functions and graphs.

The majority of the mentors participated in the summer session. The summer component consisted of twelve full-day workshops. During the morning, mentors observed lessons with students and activities modeled by the district expert. These lessons exemplified best practice.

The role of the teacher participants in these sessions was not passive. They invented word problems and interacted with students to provide learning opportunities. After the morning lessons, the mentors, curriculum facilitator, and students held debriefing sessions. Mentors analyzed elements of the lesson that were effective, improvements that could be made, and changes they could incorporate to make the lessons better fit their own styles and the needs of their students. In the afternoon, mentors formed small groups to discuss a daily focal question and other issues. Later, they engaged in activities using the same hands-on materials and addressing the same topics as the students but demanding a deeper level of understanding.

The support for mentor-teachers continued during the school year. Four-day sessions were conducted the next fall and spring following a similar format as the summer sessions. In addition, mentors met monthly. The mentors improved their own practice, provided leadership and support to other teachers in the district, and implemented standards-based curricula in the middle grades.

Positive Changes

Participants learned teaching methods that address national and state standards and that are based on research and best practice. These changes have been integrated into daily practice and have been sustained over time. Mentor-teachers have strengthened their understanding of content and the most effective methods for helping students learn mathematics. They are also more reflective about their own teaching. Further, participants now work with other teachers in their schools to disseminate their new knowledge. They have contributed to the district's overall improvement efforts as part of leadership teams. By building participants' profound understanding of the mathematics they teach, as well as expanding their pedagogical content knowledge, the mentor development program has contributed to sustained change in the district.

—Alfinio Flores and Cheryl A. Thomas

Learning Never Ends: Meeting Mentors' Professional Development Needs

Mentors come in all ages and sizes and with backgrounds as colorful and diverse as the signs in Times Square. They have no single defining characteristic, but a description that comes close is *experienced*. Not surprisingly, districts often assign more experienced teachers to serve as mentors or coaches and ask these veterans to provide professional development support to novice teachers. Although mentors are experienced, they continue to have learning needs. A systematic approach to assessing and addressing mentors' learning needs can ensure that mentors and the teachers they guide are well equipped to face the challenges of mathematics instruction.

The "Taking Stock" Tool

The National Council of Teachers of Mathematics (NCTM) issued Standards for the Support and Development of Mathematics Teachers and Teaching in its *Professional Standards for Teaching Mathematics* (NCTM 1991). Standard 2 in this set, "Responsibilities of Schools and School Systems," states that school administrators and board members should "[provide] a support system for beginning and experienced teachers of mathematics to ensure that they grow professionally and are encouraged to remain in teaching" and should "[support] teachers in self-evaluation and in analyzing, evaluating, and improving their teaching with colleagues and supervisors" (p. 181). The form shown in figure 6.1, titled "Taking Stock of the Mentor's Professional Development Needs," allows mentors to express their objectives for professional growth and encourages districts and schools to assist in meeting those goals.

Although mentors may have a wealth of knowledge about instructional strategies and methods, they may need to hone their skills in sharing this knowledge with adult learners and helping teachers improve their practices. Again, the Taking Stock tool enables a mentor's supervisor to quickly assess the mentor's perceived knowledge and level of preparedness in three areas: content, instruction, and leadership. The tool consists of four sections:

- Section I: Mentors assess their abilities to help teachers in the areas of instruction, curriculum, and leadership and the change process.

- Section II: Mentors respond to questions about their level of confidence in their coaching skills.

- Section III: Mentors rate their desire to learn more in the areas of instruction, curriculum, and leadership.

- Section IV: Mentors express additional needs or needs they cannot easily categorize.

Alignment of the Tool with Standards and Research

The Taking Stock tool is responsive to the Standards articulated in NCTM's *Professional Standards for Teaching Mathematics* (1991). For example, the questions on instruction are related to Standard 3, "Knowing Students as Learners of Mathematics," of the Standards for the Support and Development of

Taking Stock of the Mentor's Professional Development Needs

I. *To what extent do you feel prepared to help teachers ...*

	Not at all				To a great extent
1. use a variety of research-based instructional strategies to engage all (e.g., ELL, special education) students in important mathematics?	1	2	3	4	5
2. use assessment information to modify instruction for individual students and groups of students on an ongoing basis (i.e., throughout a unit, not just at the end of a unit)?	1	2	3	4	5
3. engage students in collaborative mathematics discourse?	1	2	3	4	5
4. communicate high expectations for all students through actions and words?	1	2	3	4	5
5. identify the "big ideas," key concepts, and knowledge and skills within the mathematics curriculum?	1	2	3	4	5
6. identify assessments that correlate to the conceptual understanding required and related knowledge and skills?	1	2	3	4	5
7. plan a standards-based lesson?	1	2	3	4	5
8. apply knowledge of the stages of change to adopt new instructional practices?	1	2	3	4	5
9. anticipate, identify, and resolve conflicts that are a result of the effort to improve mathematics teaching and learning?	1	2	3	4	5
10. recognize and celebrate improvements in their practices and their success in improving student achievement in mathematics?	1	2	3	4	5
11. establish and assess their own progress toward a goal for improving instruction and student learning?	1	2	3	4	5

II. *To what extent do you feel prepared to ...*

	Not at all				To a great extent
1. systematically foster the sharing of ideas and successes between teachers in the school?	1	2	3	4	5
2. provide teachers with timely feedback that focuses specifically on the characteristics of high-quality mathematics instruction?	1	2	3	4	5

(Continued on next page)

**Fig. 6.1. Tool for assessing mentors' needs and objectives:
"Taking Stock of the Mentor's Professional Development Needs"**

III. *Using a scale of 1 to 5, with 1 being low, indicate your interest in learning more about each of the following topics:*

a. effective mathematics curriculum	1	2	3	4	5
b. high-quality mathematics instruction	1	2	3	4	5
c. standards-based instruction and lessons	1	2	3	4	5
d. helping teachers through the change process	1	2	3	4	5
e. providing feedback to teachers	1	2	3	4	5

IV. *What problems do you anticipate will arise as you help teachers fully implement standards-based lessons and teaching? What might you need to learn more about in order to address these problems?*

What are two or three things you would like to learn next year to increase your ability to help teachers design and deliver high-quality, standards-based mathematics lessons?

(Denver, Colo.: Mid-continent Research for Education and Learning, 2007). Reprinted by permission of McRel.

Fig. 6.1. Tool for assessing mentors' needs and objectives:
"Taking Stock of the Mentor's Professional Development Needs"—*Continued*

Mathematics Teachers and Teaching. This Standard reads as follows:

> The … continuing education of teachers of mathematics should provide multiple perspectives on students as learners of mathematics by developing teachers' knowledge of research on how students learn mathematics; the effects of students' age, abilities, interests, and experiences on learning mathematics; the influences of students' linguistic, ethnic, racial, and socioeconomic backgrounds and gender on learning mathematics; [and] ways to affirm and support full participation and continued study of mathematics by all students. (p. 144)

The questions on the Taking Stock form are also relevant to Standard 4 of the same set of standards, "Knowing Mathematical Pedagogy." Although mentors usually excel in mathematical pedagogy, this Standard divides this topic into important concepts that are essential to high-quality teaching. By addressing these concepts individually, the Standard helps mentors reflect on their own teaching and, thus, conceive ways to mentor others more effectively.

The Taking Stock tool is based in part on research findings by Weiss, Pasley, Smith, Banilower, and Heck (2003) indicating that only a small percent of mathematics lessons are of high quality; in fact, only 17 percent of elementary school lessons, 7 percent of middle school lessons, and 12 percent of high school lessons were found to be of high quality. Several of the questions related to curriculum and instruction on this form were formulated from the researchers' findings regarding the characteristics of high-quality mathematics lessons.

Other questions were derived from reports of the National Research Council titled *How Students Learn: History, Mathematics, and Science in the Classroom* (2005) and *How People Learn: Brain, Mind, Experience, and School* (2000) and from previous research on effective classroom instruction performed by Mid-continent Research for Education and Learning (McREL; see *Classroom Instruction That Works* [Marzano, Pickering, and Pollock 2001]). Finally, because mentors need to know not only the characteristics of high-quality lessons but also the methods they can use to help teachers change their classroom practices, the tool includes questions drawn from McREL's ongoing work on leadership and change.

Use of the Tool in School Districts

District-level coordinators can use the Taking Stock tool with mentors to determine the type and content of professional development opportunities their mentors need. Unsatisfactory results for a particular set of questions may highlight the need to address that area when investing in professional development. By using a tool that inventories staff needs, districts and schools can focus their professional development efforts more accurately and make better use of the vast experience their mentors have to share.

—Heather A. Martindill

REFERENCES

Marzano, Robert J., Debra J. Pickering, and Jane E. Pollock. *Classroom Instruction That Works: Research-Based Strategies for Increasing Student Achievement.* Alexandria, Va.: Association for Supervision and Curriculum Development, 2001.

Mid-continent Research for Education and Learning. "Taking Stock of the Mentor's Professional Development Needs." Denver, Colo.: Mid-continent Research for Education and Learning, 2007.

National Council of Teachers of Mathematics (NCTM). *Professional Standards for Teaching Mathematics.* Reston, Va.: NCTM, 1991.

National Research Council. *How People Learn: Brain, Mind, Experience, and School.* Washington, D.C.: National Academy Press, 2000.

———. *How Students Learn: History, Mathematics, and Science in the Classroom.* Washington, D.C.: National Academy Press, 2005.

Weiss, Iris R., Joan D. Pasley, P. Sean Smith, Eric R. Banilower, and Daniel J. Heck. *Looking inside the Classroom: A Study of K–12 Mathematics and Science Education in the United States.* Chapel Hill, N.C.: Horizon Research, 2003.

Content-Based Mentoring

Research continues to show the importance of both content and pedagogical knowledge for effective teaching practice. Teaching standards and the accountability measures of effective teaching consistently include criteria that address content knowledge and effective instruction of that content for student learning. Given the findings of research and the criteria for teaching accountability, our profession must ensure continuity of support to assist teachers in the transfer of content and pedagogical knowledge to student achievement and teaching practice.

Effects of Content-Based Mentoring

One effective way to facilitate the process of teacher growth is through a high-quality mentoring program that incorporates content-based coaching into a larger support system for teachers. Teachers are more likely to grow in teaching practices when they are engaged in ongoing inquiry and reflection related to research-based practices in multiple aspects of teaching and learning. Teachers are also more responsive to analyzing the causes and effects of their instructional practices and content delivery when systems are in place to enhance content and pedagogical knowledge and when such systems are unconnected with professional evaluation. This professional development must align with both teacher needs and the demands of the profession.

Structures for Content-Based Mentoring

Content-specific mentoring support may incorporate a number of different structures, as described below.

- As part of professional development sessions at the school or district level, breakout sessions with content experts may be offered to address content-specific challenges and needs for educators. Teachers engaged in these extensions of learning report a higher degree of relevance when they apply their learning to specific teaching contexts. For example, a breakout session for mathematics teachers in a seminar on classroom management might address the use of materials, resources, and manipulatives and procedures for group work. In a seminar on instruction, the focus group might address the challenge of promoting student inquiry specific to mathematics concepts.

- A one-to-one mentor may be assigned who teaches in the same content area as a mentee. This mentor should be trained in mentoring skills to facilitate relationship building, conduct needs assessments to plan for individualized growth, and engage in meaningful content-based activities to support lesson planning and student assessment. Ideally, the mentor should work at the same school site and have experience teaching a grade as close as possible to that taught by the assigned mentee.

- An instructional coach may observe classroom lessons, coteach, and hold prelesson and post-lesson conferences with an assigned teacher. The coach may be either a generalist or a content-based expert. A generalist coach might perform a needs assessment with the teacher, provide mentoring support aligned with teaching standards, and serve as the liaison to bring in specialists when the teacher needs specific support in his or her content area. The coach may also refer the teacher to a content-based mentor. A content-specialist coach provides content-based mentoring and support for standards-based teaching.

- A process may be put in place to ensure individual reflection and goal setting, which in turn facilitate improvements in teaching. The simple act of looking back at one's teaching and planning for the future leads to sustained change over time. Questions that prompt meaningful reflection for the teacher may include the following: What aspects of my teaching positively influenced student learning? What evidence do I have that students learned and can apply content? What would I do differently next time to predict a higher success rate for my students? Focusing on their specific teaching and learning environments makes the process of reflection more meaningful for teachers.

Conclusion

The question for our profession is not whether to adopt content-based support but how to integrate such support into a system in which teachers embrace the growth associated with content knowledge, content de-

livery, and assessment of student learning. The support of a comprehensive mentoring program allows teachers to view themselves as members of a professional learning community and helps them increase their confidence and competence in teaching.

—Sharon A. Kortman

Mentoring for High-Quality Instruction Using Adult Learning Theory: Lessons from Research and Practice

Every good teacher plans for class by preparing introductions to ignite students' interest in the topic of the day, by developing activities to engage students with content, and by formulating essential questions to push students' thinking to a higher level. Mentors are good teachers who also have the knowledge and skills to assist their colleagues. Many mentors have received extensive training in instructional processes and pedagogy, that is, "the art and science of teaching [children]." But because mentors work with adult colleagues, they may benefit from a greater understanding of *andragogy*, "the art and science of helping adults learn" (Knowles 1980, p. 43).

Basic Assumptions of Andragogy

At the heart of andragogy are the following five major assumptions about adults as learners (Knowles et al. 1984; Knowles 1990):

1. The self-concept of adults is less dependent on the opinions of others and more internally motivated than that of children.

2. Adults have a wealth of experience that can enrich learning.

3. Adults' readiness to learn is closely related to skills necessary in their social roles.

4. Adults are more responsive to problem-centered learning than subject-centered learning.

5. Adults are motivated to learn by internal rather than external factors.

Various corporations, government organizations, and educational institutions apply these assumptions to their employee training. Researchers also agree that the assumptions of andragogy can be classified as principles of good practice (Merriam and Caffarella 1999).

As Elmore (2000) notes,

> … the job of the administrative leaders is primarily about enhancing the skills and knowledge of people in the organization, creating a common culture of expectations around the use of those skills and knowledge, holding the various pieces of the organization together in a productive relationship with each other, and holding the individuals accountable for their contributions to the collective result. (p. 15)

Elmore's reference to administrators is applicable to the mentoring relationship. Mentors should focus on enhancing the skills and knowledge of their protégés while accounting for the specific characteristics and objectives of adult learners.

Suggestions for Practice

Applying the concepts of andragogy to collegial coaching is both easy and beneficial to those in the mentoring relationship. Below are some suggestions for putting the concepts of andragogy into practice, followed by brief explanations of the benefits gained.

- Maintain open communication by allowing the mentee to verbalize his or her achievements and problems while the mentor listens actively. By engaging in active listening, the mentor may be able to identify core problems. Mentors should restate problems they identify to clarify any misinterpretations. Working together to develop strategies to solve problems builds trust in the relationship. Mentors should also keep in mind that starting with smaller issues creates early success.

- Have an open discussion to set goals and expectations for the mentee, but avoid identifying too many goals. Having three primary goals in one school year is sufficient. Because mentors usually work with new teachers, they should keep in mind that their protégés may be overwhelmed by the process of acclimation to the system, the school, the students, and daily planning. The mentor should also be certain that the goals chosen are related to problems identified by the teacher. Additionally, mentors should incorporate goals for themselves. The mentor's goals can be simple, such as holding both mentor and mentee responsible for meeting expectations, providing resources for the teacher, or modeling deep, honest reflection.

- Have both parties agree to keep a journal with daily or weekly entries on conversations, thoughts, feelings, problems, strategies, and solutions related to teaching. The activity may have more depth and meaning if the mentor and mentee develop two or three essential questions to focus their writing, with the understanding that additional information may be added. Questions for the teacher might include these: What was my biggest struggle this week? Am I using my time effectively? Are my students engaged in my lessons, and how do I measure engagement? The mentor's questions might include the following: Do I actively listen to the concerns of my protégé? Do I allow my protégé to find his or her own solutions to problems? Semimonthly "journal swapping" allows both parties to gain a sense of the other person's thoughts and positions on important issues. Many people are guarded when speaking but tend to open up more in journal writing.

- Maintain the focus of the relationship by holding monthly meetings to reflect on initial goals and problems. If all the goals have been accomplished, begin the process again, identifying new objectives. Planning, implementation, and evaluation are elements of a collective, cyclic process. If one element is left out or the cycle is not repeated often enough, the whole experience is diminished.

Final Thoughts

Reflective practitioners are central to the process of mentoring. Serving as a mentor, coach, or administrator is an honor that allows one to learn more about pedagogy, get to know colleagues, and gain a greater understanding of the learning process. Through open communication and reflection, mentors not only assist their colleagues but also improve their own practice. Adult learning theory and practical experiences may serve as tools to be applied to reflection, learning, and growth for both mentors and protégés.

—Thomas J. Starmack

REFERENCES

Elmore, Richard F. *Building a New Structure for School Leadership.* Washington, D.C.: Albert Shanker Institute, 2000.

Knowles, Malcom S. *The Modern Practice of Adult Education: From Pedagogy to Andragogy.* 2d ed. New York: Cambridge Books, 1980.

————. *The Adult Learner: A Neglected Species.* 4th ed. Houston, Tex.: Gulf Publications Co., 1990.

Knowles, Malcom S., and Associates. *Andragogy in Action: Applying Modern Principles of Adult Learning.* The Jossey-Bass Higher Education Series. San Francisco: Jossey-Bass, 1984.

Merriam, Sharan B., and Rosemary S. Caffarella. *Learning in Adulthood: A Comprehensive Guide.* 2d ed. San Francisco: Jossey-Bass, 1999.

Collaborative Mentoring: Establishing a Mathematics Teaching and Learning Community through Lesson Study

Collaborative mentoring empowers mentors by redefining the tasks they perform as a collaborative process of continual professional learning. The collaborative mentoring approach combines lesson study (Lewis 2002) and the model of professional learning communities (DuFour 2004) to provide opportunities for new and experienced teachers to engage in shared professional learning. While working collaboratively with the grade-level teams, the mentors described here capitalized on the shared expertise among a group that included novice and experienced teachers, special educators, and teachers of English-language learners. The exchange of mathematical ideas and instructional strategies that took place among these teachers helped create a sustainable, teacher-led professional learning community.

The mentoring process involved three phases: (1) collaborative planning, in which novice, experienced, and special educators and mentors worked together to plan lessons; (2) teaching and observation, in which one teacher taught the focus lesson and others observed using an observation protocol; and (3) debriefing, in which teachers reflected on the lesson design, task, and student engagement and learning and discussed future

steps. Some of the guiding questions for the teaching and learning processes were as follows:

- What is the important mathematical understanding that students need to learn?

- What are potential barriers and anticipated student responses?

- What conceptual supports and instructional strategies can best address our students' learning? How will we respond when students have difficulty?

- How will we know when each student has learned the concepts taught in the lesson?

Three Discoveries about the Collaborative Mentoring Process

Perhaps the most important discovery by the collaborative mentoring group was that the extent and value of shared learning was greater than many members anticipated. The reciprocal learning relationship among the novice teachers, experienced teachers, special educators, and teacher-educators was evident in the discourse that took place during the planning phase. As each teacher contributed to the group knowledge, different levels of mentoring and expertise were revealed. For example, novice teachers were mentored by experienced teachers and special educators who had specific knowledge of potential barriers to learning, common misconceptions, and likely student responses acquired through their years of experience working with diverse student populations and through special training. In turn, novice teachers who were recent graduates or were enrolled in master's degree programs shared their knowledge of new strategies, curriculum developments, and the latest technological tools with the experienced teachers and special educators. The mathematics educator and the mathematics specialist bridged the instructional practices and strategies with supporting research on mathematics teaching and learning.

The members of the collaborative mentoring group also came to realize the importance of relearning the mathematics they were teaching. In the collaborative planning phase, teachers constructed a mathematics knowledge map that outlined prerequisite knowledge, interrelated concepts, and concepts that would serve as building blocks for future knowledge. In addition, they identified effective representations or models to teach those mathematical ideas. One teacher reported,

> The mapping of prior knowledge needed and future knowledge was illuminating—it just got me thinking more deeply about the concept. The brainstorming helped to see what kids need to know and where they are headed. It makes it easy to see all of the standards that are tied into one concept. I learned about multiple models of representations and strategies.

This relearning process allowed new and experienced teachers to further build their mathematical knowledge in terms of concepts, models, strategies, and representations.

Finally, the collaborative mentoring approach established for participating teachers a professional learning community for the teaching of mathematics in which teachers openly shared instructional practices. As one teacher wrote in her reflection, "[The mentoring approach was an] excellent way to open communication between teachers. Lesson study is a great model for team planning. Many brains are better than one. The exchange between these different teachers was enlightening."

Conclusion

Inclusive classrooms and coteaching approaches have not prompted the shift to collaboration. Collaboration has supported inclusive classrooms and coteaching. Teachers learn to draw on the strengths and specialties of others and continue to enhance their practice by identifying, implementing, and refining instructional resources and strategies through coaching, coteaching, and lesson modeling. In the collaborative model, mentoring is not associated with a separate induction process but is seen as part of continual professional learning. Rethinking the traditional model of mentoring and coaching to consider the collaborative mentoring approach may optimize the professional development and improve the mathematical knowledge of both new and experienced teachers.

—Jennifer Suh and Spencer Jamieson

REFERENCES

DuFour, Richard. "Schools as Learning Communities." *Educational Leadership* 61, no. 8 (May 2004): 6–11.

Lewis, Catherine C. *Lesson Study: A Handbook of Teacher-Led Instructional Change.* Hillsdale, N.J.: Research for Better Schools, 2002.

Essential Components of a Novice Teacher Induction Program

Leslie Huling and Virginia Resta served as the principal investigators for a $4.7 million grant from Houston Endowment, Incorporated, to develop a new teacher induction program (NTIP). The five essential components of the program provide a support system that goes beyond teaching strategies and curriculum development to address the realities faced by a new generation of novice mathematics teachers with unique needs and concerns. These five components are described in the following sections.

University Faculty Participation and Graduate-Level Classes

As part of the NTIP, participants completed six hours of tuition-free, graduate-level coursework. Mentees, university faculty members, and mentors met every other week to discuss topics of interest, such as classroom management, students with special needs, mathematics resource materials, and innovative instructional strategies that keep students motivated. The mentors and university faculty surveyed the first-year teachers to discover their needs and, at times, invited specialists to the meetings, for example, the district mathematics coordinator, who explained the curriculum. These three-hour graduate-level sessions helped build camaraderie among the new teachers in each school district. In some instances, the university professors taught classes off campus for the convenience of the mentees.

Collegial Atmosphere among Mentees

Not only did the mentees enjoy getting together every other week in class to share stories and exchange ideas, but they also kept in touch at other times. As part of the course requirements, these new teachers logged on to NiceNet, a free, Web-based learning environment. The friendly chat format of this site helped mentees feel at ease in asking questions.

Assignment of Individual Mentors

Most mentors in the program were retirees who were paid approximately $24,000 yearly to visit ten to twelve NTIP participants each semester. Weekly mentee visits ranged from quick stops to much longer meetings, depending on the new teacher's particular needs. Mentors assisted with a range of tasks, from modeling mathematics lessons to locating mathematics resource materials for English-language learners.

At times, the mentors became sounding boards while the new teachers vented their frustrations. As part of their role, mentors served as listeners, counselors, advisers, cheerleaders, confidants, and supporters. Bimonthly meetings and yearly mentor conferences helped the mentors gain new insight into their roles and allowed them time to share ideas. Mentors also had access to a NiceNet section where they could pose questions, seek advice, and tell success stories.

Cooperation of District and Campus Staff

Principals and counselors from each campus became role models to help novice teachers learn the habits, routines, and behaviors that reflected the beliefs, norms, and values of their particular schools. This support helped minimize conflicts between mentees' educational views and school expectations. Participating principals were informed about NTIP concepts ahead of time, demonstrated their support of the program, and agreed to abide by program expectations.

Access to Shared Teaching Resources

University professors, mentors, and mentees all shared teaching ideas and resources, such as books, journal articles, and Web pages, with other program participants. As part of the course requirements, the mentees created a *Bright Ideas* book that included their best lesson plans and proved to be an invaluable resource.

Impact of the Five Components

With these five components in place, novice teachers excelled and were retained in the teaching profession. The principal investigators currently track all NTIP teachers for five years, and the results to date indicate that more than 92 percent have remained in the profession. As one third-grade teacher, a mentee in the program, commented, "Keep helping those first-year teachers because it makes such a big difference."

—Patricia A. Williams, Sylvia R. Taube, and Margaret A. Hammer

Challenges and Resolutions for Mentoring Teachers of Mathematics

The following sections describe challenges and resolutions to address those challenges in the mentoring of teachers of mathematics.

Mentoring Structure with a System Perspective

Challenge: *The numbers of mentors and instructional coaches employed by school systems have increased, but these professionals lack a clearly defined support system.*

As the prevalence of mentoring and coaching grows, some school districts have increased the number of instructional coaches they employ without creating an infrastructure to maximize the time and resources of these coaches. Those who suffer in this scenario are the mentees, who are sometimes overwhelmed by the variety of support they receive. A new teacher may be confused, for example, when literacy, mathematics, and technology coaches all show up on the classroom doorstep.

Resolution: *Adopt a parallel mentoring model with an on-site mentor and a district- or site-based instructional coach.*

In this model, a mentor who works close to the same grade level and in the same content area as the mentee provides daily on-site support. The role is clearly defined and offers benefits to the school community. The parallel mentoring comes from an instructional coach, who observes lessons in the classroom, offers guidance that focuses on instruction, and provides content- and standards-based coaching. The coach also serves as a liaison to introduce additional support when the teacher is developmentally ready for it. The teacher views the layered support as comprehensive and coordinated.

Mentors Equipped with Skills to Effect Change in Teaching Practice

Challenge: *New teachers receive content support from unskilled mentors or relational mentoring void of instructional and content-based coaching.*

In either of these scenarios, the mentee loses out on the growth and development that can occur through a powerful mentoring model that incorporates instructional and content-based support in the context of a collegial mentoring relationship.

Resolution: *Provide appropriate mentor training.*

Whether a mentor is a content-specific coach or a generalist, he or she needs to understand the mentee's complex set of responsibilities to address the totality of the teaching role. The most effective mentors are content experts who are trained in mentoring skills or generalist mentors who are trained in content-specific questioning strategies and processes. More often than not, effective mentors are not born but made. A proactive training model for mentors has hidden benefits of teacher growth at all levels of development and the establishment of professional learning communities. In turn, these benefits have a positive impact on the entire school system. As mentors learn the skills of their job, they develop trusting relationships with new teachers, accomplish joint work with their colleagues, and analyze their own practices. With training, mentors learn to provide appropriate content-based support, analyze teaching standards, view students' work, measure teachers' growth, and assess studens't progress. In other words, training for mentors translates to achievement of expected outcomes in teachers' growth.

Mentoring as the Link between Teaching Practice and Student Learning

Challenge: *Mentoring focuses too intently on the teacher at the expense of the students.*

Sometimes, well-intentioned mentoring programs are so focused on providing support to teachers that intentional links with students and their achievement become secondary.

Resolution: *Ensure application of teaching knowledge and skills to students' learning and success through mentoring.*

Research shows that effective professional development increases teacher competence, and as we know, the quality of teaching is directly linked with students' success. Mentoring as a professional development model, then, must reinforce the link between the quality of teaching and students' achievement. The fundamental idea here is to transfer our knowledge of students' learning to the mentoring relationship. We know, for example, that positive reinforcement of behavior leads to repetition of that behavior. In the mentoring relationship, mentors should celebrate successful teaching practices of their mentees one skill at a time. We also know that the more specific feedback students receive on their learning, the more likely they are to apply that feedback to increase their knowledge or skills in the future. In working with mentees, mentors should align the causes and effects of instructional practices with students' learning, management, and assessment. This approach can be applied to both the strengths of, and areas in need of refinement by, mentees. The more explicitly these factors are linked, the more likely the lessons learned are to be applied to teaching practice. By transferring our knowledge of students' learning to new teachers, we can help them make significant gains in their professional practice.

—Sharon A. Kortman

Section 7: Lessons Learned

THE MAIN focus of this book is on collegial and collaborative relationships between professionals who are dedicated to, and passionate about, the teaching and learning of mathematics. At one time, teaching was an endeavor that took place behind closed doors, with each teacher responsible for figuring out what was best for the students in his or her classroom. Teachers did not engage their colleagues in discussions about what worked or did not work in practice. Admittedly, some teachers have more opportunities for collaboration than others, but no one can deny the value of the learning that takes place between mentor and protégé—the protégé learning from the mentor and, often, the mentor learning from the protégé.

This last section, then, discusses "lessons learned" in the mentoring process. Who knows better than a teacher that life is about learning from our mistakes? Yet we can also learn from the successes as well as the mistakes of others, those who have traveled down the same road and are willing to share their insights. The authors in this section have performed precisely that service. Realizing that each of our experiences is unique, we cannot anticipate every situation or concern that will arise; however, if the advice offered here can improve the mentoring relationship even minimally, both the mentor and protégé will appreciate the experience all the more.

66 Experience is a hard teacher because she gives the test first, the lesson afterwards. 99

—*Vernon Sanders Law*

What I Wish I Had Known: Mentoring the First-Year Teacher

> Experience is a hard teacher because she gives the test first, the lesson afterwards.
>
> —*Vernon Sanders Law*

My mother, a career teacher, once told me that not all knowledge comes from a book. At the time, I thought this statement, coming from a woman devoted to learning, was rather strange. After my first year of teaching, however, I understood what she meant. Some of the lessons I learned in that first year could not be found in a textbook. No degree of training can prepare a new teacher for walking into a classroom for the first time.

The Innocence of Youth

I had envisioned spending my first week teaching compelling and original lessons that would bring the study of mathematics alive for my students. Instead, I spent most of my first week as a teacher learning how to navigate the halls from one classroom to the next; how to work the copier, my phone, and my computer; and how to take attendance, check homework, teach, and manage the discipline of twenty students in a fifty-five-minute period. At the end of the day, I would count my blessings if I had arrived in each classroom from my office across the building with everything I needed for the day's lesson. The first few months of my first year were challenging. In those months, I relied on the experience and advice of the veteran teachers in my school and the instructors in my teacher education program. Sometimes, their counsel saved me from learning a lesson the hard way, and when it did not, their support was invaluable. As the months wore on, I began to feel more confident, and by the end of the year, if I did not know the answer to one of my questions, I knew whom to ask.

The Lessons of Experience

In the final meeting with the teachers in my teacher education program, our instructor asked us a question I will never forget: "What do you wish you had known at the beginning of this year?" After almost an hour, we new teachers still were not finished compiling our lists. The instructor's question was a good one because it required us to reflect on the mistakes we had made during the previous year and the lessons we had learned from those mistakes. I was a better teacher in my second year because I answered that question. Below are some of the lessons I learned from my first year of teaching.

- Learn the practical aspects of how your school works. Find out who is responsible for getting your computer and the copier fixed. Ask how to go about purchasing supplies you need for your classroom. Get to know the school support staff and their duties so that you know how to get help when you need it.

- Make a copy of any forms you have to fill out. That way, when you have to fill out the form again, you have the copy to use as a guide.

- Choose four or five achievable goals for the year. Narrowing your focus in this way will prevent you from feeling overwhelmed. For your second year, pick four or five different goals.

- Know the priorities set by the principal and your department. This understanding will allow you to set realistic goals that are in line with school expectations.

- Set reasonable limits for how much time you spend at school or working from home. Give yourself one afternoon a week off from tutorials. Share the monitoring of students serving detention with another teacher. Make parent phone calls only from school. Balancing work life and home life will nurture you as a person and make you a better teacher.

- Meet regularly with a more experienced teacher in your content area to ask questions about the lessons: With what concepts do students typically encounter problems? Do the students have difficulty with any of the problems in the homework assignments?

- Keep a daily record of behavior problems, and track each student's completion of homework. This practice may help justify a strict behavioral penalty or a failing grade to parents and school administrators. The best defense is a good offense in both teaching and sports.

- Buy any textbooks for your grade level or subject that you can find, no matter how old they are. Such books are great resources and will give you ideas for potential assessment questions or examples to present during your lessons.

- Write the first test or quiz you give with a more experienced teacher. Otherwise, it may be too short, too long, too easy, or too hard.

- Write the assessments before you write the lesson plan. The lesson should include examples with the same vocabulary or wording as the questions on the homework and assessments.

- Never make an assignment that you are not willing to grade.

The Transmission of Hard-Earned Wisdom

This year, I find myself on the other side of the mentoring equation, serving as a supervisor of teachers in a teacher education program at a university. I hope that the advice in my list will prevent my mentee from learning a lesson or two the hard way. If not, at least the list serves as a reminder that the first year of teaching comes with a steep learning curve and that all experienced teachers were once first-year teachers, too. At our last meeting at the end of this year, I will have just one question for my mentee: "What do you wish you had known at the beginning of this year?"

—*H. Smith Risser*

REFERENCE

Law, Vernon Sanders. Quotes.net. STANDS4 LLC, 2008. http://www.quotes.net/quote/20785.

Difference of opinion leads to inquiry, and inquiry to the truth.

—*Thomas Jefferson*

At first glance, the title of this article may seem counter to common sense. Minimizing or eliminating differences of opinion[1] might seem more appropriate than capitalizing on them. Differences of opinion can be unpleasant experiences that individuals seek to avoid, but they can also serve as productive opportunities for learning among teachers. Expressing differences of opinion leads individuals to examine the rationale for the positions they hold, and this self-examination and reflection are essential elements of teachers' growth. When properly managed, differences of opinion between a mentor and a new teacher can become learning opportunities rather than unpleasant confrontations. As an illustration of the potential value of differences of opinion, consider a vignette about a first-year teacher, Ellen,[2] and her mentor, Ben.

Ellen's Lesson

Ellen was concluding a first-year-algebra unit on linear relationships in her class. She had allocated one day of instructional time for review before the unit test. To design a review activity, Ellen used a software program she had learned about in an educational technology course that allowed her to create a Jeopardy-type game for students to play. Ellen spent a great deal of time writing questions to put into the program. She also tracked down an LCD projector and a laptop computer from the school library so that the game board could be projected on the screen at the front of the classroom. Ellen was so excited about the lesson that she asked Ben to observe it.

---

1. The introductory quote by Thomas Jefferson about differences of opinion can be found in "Constructive Controversy: The Value of Intellectual Opposition" by David W. Johnson, Roger T. Johnson, and Dean Tjosvold in *The Handbook of Conflict Resolution,* 2nd ed., edited by Morton Deutsch, Peter T. Coleman, and Eric C. Marcus (San Francisco: John Wiley & Sons, 2006, pp. 69–91).

2. All names in the article are pseudonyms.

After seeing Ellen's enthusiasm about the lesson, Ben agreed to observe, but he had some misgivings about the review activity. Ben believed that playing Jeopardy and similar games disrupted the learning process. He thought that such games rewarded students who were able to shout out answers quickly but not those who thought deliberately. Students who wanted time to look at problems from many angles were discouraged from doing so because of the rapid pace of the game. In his estimation, Jeopardy and similar games frequently led to two types of inappropriate student responses: aggressiveness from students who were skilled at quick recall of facts and indifference from students who were not. Nonetheless, Ben did not discourage Ellen from conducting the lesson because he saw her excitement. He told himself to go into the lesson with an open mind and be attentive to its positive aspects.

Ben's Observation

Unfortunately, some of Ben's initial misgivings about the lesson seemed to be confirmed by his observations. Although students worked in teams of five to construct answers to the game questions, most of the work was done by one or two members of each group. The other group members were either uninterested in the problems or absorbed with tracking the progress of the other groups. As soon as one of the groups finished, the group leader shouted out the answer and the other groups stopped working on the problem to move on to the next one. Ben also noticed that Ellen's game questions focused mostly on procedural knowledge, such as solving for x in the equation $4x + 7 = 5$ or finding the slope and y-intercept of $y = 3x + 5$. As the lesson unfolded, Ben puzzled over how he could share his observations with Ellen without discouraging her.

Postlesson Conversations

Ellen and Ben talked about the lesson immediately afterward during their shared preparation period. Ellen, still excited about the lesson, asked Ben for his thoughts. A look of disappointment crossed her face as Ben shared his observations about some students' lack of engagement in the lesson and the absence of questions that focused on conceptual understanding. Ellen fell silent after Ben finished speaking, briefly thanked him for his

observations, and excused herself to make some copies for her next class period. Given that the two teachers had always had pleasant encounters, Ben was surprised at Ellen's quick departure. He decided to approach her again the next day after the dust had settled.

The next day during the shared preparation period, Ben approached Ellen and apologized for the tone of his feedback from the previous day, which may have seemed harsh. He explained that he wanted to see her succeed as a teacher and that he would have been remiss if he had not mentioned parts of the lesson that he thought needed work. Ellen also apologized for disengaging so quickly from their conversation the previous day. She explained that she was initially defensive because from her perspective, the lesson had been successful. Overnight, she had taken some time to reflect on his comments and realized the need to restructure the lesson to improve students' engagement and conceptual thinking. Together, Ben and Ellen reworked the lesson to attain these goals. In the process, Ellen gained a more polished lesson and Ben acquired deeper knowledge of some of the technological resources available at the school. Before working with Ellen, Ben believed that the benefits of using technology were outweighed by the time required to set up and learn the hardware and software. Both individuals benefited from a situation that initially involved an uncomfortable difference of opinion.

Differences of Opinion as Learning Opportunities

Although Ben and Ellen eventually capitalized on their difference of opinion in this vignette, their work could well have been sidetracked by their unpleasant exchange immediately after the lesson. To avoid such awkward situations, mentors must establish a rule that disagreements and differences of opinion will be normal occurrences in pedagogical conversations and lesson debriefings. Complete agreement will not occur in every situation, and consensus is not necessarily the goal. Differences of opinion offer the mentor and the new teacher opportunities to grow as they push each other to consider new perspectives on classroom practices.

—Randall E. Groth

Challenges and Suggestions for Cross-Cultural Mentors

During a course in mathematics methods taught by Fatma, a university instructor, preservice teachers engaged in a discussion about the use of children's songs in number-sense instruction. Fatma encouraged the preservice teachers to sing several songs, such as "One, Two, Buckle My Shoe" and "Ten in a Bed," and discuss how to use them in class with young students. Fatma commented, "I do not know these songs because I was raised in Turkey. I grew up with similar songs in my own language. Let's hear these songs from you." As she was walking around the classroom, one of her students softly said, "You are not an American if you don't know these songs!" The student's voice was just loud enough to ensure that Fatma picked up the comment but low enough that the other students did not hear it. The discussion continued, and Fatma acted as if she had not heard the student's remark. This insensitive act on the part of the student highlights the needs of those who mentor across cultures, particularly mentors who are members of a minority culture in a given context.

One of the first challenges international mentors usually face is a language difference. As nonnative speakers of English, we have sometimes borrowed teaching materials from colleagues who are native speakers, only to be told by students that the materials contain minor writing errors. The students are quick to blame these errors on the fact that English is a second language for us, but the same students ignore the errors when they are presented by our colleagues who are native speakers of English. Cross-cultural mentors should be aware of the possibility that some mentees may be more critical of errors in communication when they are made by nonnative speakers.

Below are some suggestions to help mentors who have language or cultural differences from their mentees, particularly when the mentors are part of a minority culture.

- Be aware of the nuances of verbal and nonverbal communication in the setting in which mentoring or instruction will take place.
- Seek help from leaders at the school, native-speaking colleagues, or other professionals who have worked with similar groups of preservice teachers or mentees.
- Examine the meaning of mentoring in your culture before trying to understand mentees' needs, and encourage your mentees to do the same. Both parties should have a clear understanding about what to expect from the relationship.
- Provide mentees with a clear definition of their roles.
- Practice active listening, and provide continual feedback. These techniques help both parties in the relationship understand each other's cultures and perceptions of mentoring. In particular, mentors may ask mentees to complete reflective journal assignments and mentoring process evaluations.
- Work to establish trust with mentees so that they feel free to reflect honestly on the learning process.
- Let mentees hear you speaking in their native language. This strategy may eliminate the mystique associated with a second language and trigger mentees to appreciate your efforts to use their language.
- Do not become frustrated by cultural differences and cross-cultural mentoring challenges. Try to address and overcome your mentees' stereotypes by building trustworthy and caring relationships.

—*Fatma Aslan-Tutak and Adem Ekmekci*

Learning from a Novice Mentor's Mistakes

Years ago, during my second year as a high school mathematics teacher, I was asked to allow another teacher's student-teacher, Mary[1], to teach one of my classes. I gave Mary the option of choosing which class to teach, and she requested my last-period first-year-algebra class. I warned her that this class was a tough one and advised her to choose another period of the day, but

---

1. All names in the article are pseudonyms.

she insisted. I relented, as though my hands were tied. Mary started teaching my class a few days later and had difficulty with the students. I did not really know how to help her and, to my regret, basically left her to her own demise. In the end, Mary became something of a sacrificial lamb—she had a terrible experience and so did my students. I, on the other hand, returned to a group of students who were happier to have me as their teacher than they were before Mary showed up.

I learned two important lessons from this experience. First, I was completely unprepared to be Mary's mentor, and this deficiency had a significant negative impact on the experience for Mary, my students, and me. My lack of knowledge and training caused me to miss what could have been a valuable learning experience for all of us. Those who organize field experiences must ensure that mentors are equipped in all ways for their important role.

I also realized later that I had "used" Mary in some ways for my own selfish purposes. I wrote her off as a lost cause and waited in the wings, knowing that her experience would be negative and that my slight advantage in terms of teaching skill would cause me to look better when she left. I had the chance to reclaim my algebra students after their encounter with student-teaching, but I was unable to help Mary recover from her experience with my students. I wish I had been better prepared to support Mary, but more important, I wish I had given her the time and respect she deserved as a student, assistant, resource, and future colleague. I hope Mary has since forgiven me and that other novice mentors might learn from my mistakes.

—Keith R. Leatham

Mentoring Alternative Entrants

All new teachers may benefit from mentoring, but some require a greater level of attention because of their backgrounds. *Alternative entrants* (AEs) do not hold education degrees but are considered to have the content background needed to teach mathematics. They may be recent college graduates or individuals changing careers after several years of work in another field. AEs are typi-

cally hired with temporary teaching certificates and are given several years to meet pedagogical requirements to become fully certified. AEs are unique because they possess characteristics of both preservice and in-service teachers. Like preservice teachers, their only formal experience with schools comes from their time spent as students, yet they function as in-service teachers while they earn their credentials and learn to teach effectively. Unlike in-service teachers, however, AEs have little or no previous experience as mathematics instructors. From our collective experiences working with AEs, we have identified issues unique to this population. The following sections discuss these issues and offer suggestions that mentors might use to address them.

The Language of the Profession

AEs are often inundated with unfamiliar vocabulary that is specific to the educational setting. For example, educators may use such words as *standards* and *benchmarks* or such abbreviations as *SES*, *LEP*, and *ADHD* in the course of a discussion. Unlike preservice teachers, AEs come to the profession without a background in education; thus, many of these terms are foreign to them. Mentors should help AEs make sense of the many new words that they will encounter in the school environment.

Mentors should assure AEs that they are not expected to understand the new vocabulary immediately and that this knowledge will develop over time. Mentors should also be ready to provide translations for AEs as needed. In our experience, some AEs may be reluctant to ask questions because they assume that they are expected to know the jargon and do not want to call attention to their lack of knowledge. If mentors explain terms readily or clarify information without prompting, AEs may be more open to asking questions.

Teacher Workload

Many AEs come to the profession without a full understanding of the diverse responsibilities of teachers. They may have an idealistic view of teaching drawn from their personal experiences and may not be prepared for the day-to-day realities of a mathematics classroom. For example, many AEs are surprised and overwhelmed by the amount of paperwork, including school records and homework, that they are required to handle and other time-consuming activities for which they are responsible. They might be frustrated because they are not familiar with the documentation they are expected to maintain, such as attendance records; may not have efficient procedures for dealing with instructional activities, such as collecting and returning homework; and may struggle to keep pace with the vast amount of information they are given and the number of action items they have to complete.

Mentors working with AEs should help them anticipate and develop plans for addressing periods of potentially heavy workloads. Because they lack instructional experience, AEs may not have the organizational background to help them sort and prioritize their responsibilities. When AEs appear to be under stress or reveal that they are overwhelmed, mentors should ask about their current organizational practices and help them consider alternative approaches that may be more efficient. Mentors must be careful not to prescribe particular methods but, rather, to help AEs examine possible options for handling their responsibilities.

Understanding the Schooling Process

As part of their internship experiences, preservice teachers are given varied experiences to help them understand the overall organizational structure of schools, including identifying essential personnel and the roles they play. AEs may come to teaching without this knowledge and may not be aware of the resources available to teachers at their schools or in their school districts. Mentors can help AE teachers make these connections, particularly during their first few months of school, by directing them to appropriate school or district staff members and walking them through administrative procedures. This kind of information should be shared sparingly and at appropriate times, however, to avoid information overload for the AEs.

Nature of Mathematics Teaching and Learning

Change-of-career AEs often find that the expectations for teaching and learning have changed significantly since they attended school. They may have difficulty reconciling their experiences as students with

the environment, concepts, and methods they encounter as teachers. For example, they may be unfamiliar with some mathematics topics, such as stem-and-leaf plots and tessellations, or they may struggle with concepts or procedures that they have not used since they were students. Further, some AEs may have difficulty translating and sharing their mathematics knowledge in ways that are useful to students.

Mentors should assure AEs that periods of uncertainty are expected and that both parties will work together to develop instructional skills. Mentors can assist AEs by codeveloping and analyzing lessons, modeling or coteaching lessons, and providing feedback about observed lessons. The goal is for mentors to encourage the professional development of AEs by being supportive, sharing ideas, making suggestions, and modeling effective instructional techniques.

—Gladis Kersaint, Joy B. Schackow,
Janet Boatman, Tammy Rush,
Virginia Harrell, and Joyce McClain

Building an Open Relationship: A Mentoring Vignette from Both Perspectives

Michael: When Nicole approached me with serious concerns about her future as a teacher, I knew she might be facing a defining moment in her professional development. Spending the extra time to listen to her concerns initiated a mentoring relationship that has extended for the last three years.

Nicole: I first met Michael as a sophomore in his practicum in mathematics. Outside our classroom, Michael listened to my concerns about becoming a teacher, reassured me by showing respect for my concerns and sharing his experiences, and invited me to come back with further questions or problems. Michael's advice and support shaped who I am today, helping me to discover for myself the confidence to become a teacher. Michael also continued to support me through numerous other endeavors, including earning a summer job as a geometry teaching assistant for gifted students aged twelve to fourteen—an experience so rewarding that afterward, I had little doubt that I was meant to be a teacher—and acquiring a Fulbright scholarship to teach overseas.

As partners in a relatively successful mentoring relationship, we have identified specific characteristics and goals that helped us sustain an open and honest connection. These characteristics and goals are described in the following sections.

Michael's Input

- Respond quickly to mentees' needs. Time is a precious commodity, but change may take place slowly and individually. When confronted with scheduling choices, I always make time to mentor Nicole. I try to respond promptly to e-mail messages or phone calls and, if necessary, delay work on larger projects to speak with Nicole.

- Sharpen your listening skills. Although I have not always been a good listener, I have realized the need for improvement in this area in my role as a mentor. One way to become a better listener is to summarize the mentee's ideas before moving on to your own thoughts. Focusing on mentees' ideas helps these preservice teachers make the connection between theory and their own practice. Mentors should also avoid solving problems for their mentees, instead encouraging their protégés to come up with their own ideas.

- Show that you care. Mentors should truly care about the success of their mentees and make sure that the mentees are aware of that commitment. I sometimes find myself too quick to criticize and must actively choose to give positive feedback.

- Give specific feedback. Avoid general comments; more specific feedback helps the mentee identify meaningful core concepts. For example, I once told Nicole, "Your lesson plan has some nice closure questions. They help you check for understanding about the main ideas of prime numbers, factors, and composites and simultaneously ask your students how to win this mathematics game."

- Encourage growth. Mentors should push mentees to expand their horizons before they leave their undergraduate institutions. Conferences held by the National Council of Teachers of Mathematics, undergraduate research or curriculum opportuni-

ties, and grant or scholarship awards are among several options that help mentees establish their own professional development expectations.

Nicole's Input

- Pick someone you trust as a mentor. I knew Michael's teaching philosophy and identified him as a safe and reliable adviser. You will have different opinions than your mentor, but you must have the same general philosophy and goals in mind to have a successful mentorship. This trust is a foundation for a positive, long-term mentorship.

- Do not expect easy answers. Our mentoring relationship was one of the first I experienced in college. Initially, I expected Michael to interpret my interests and concerns and tell me which road to take. When he did not tell me exactly what to do, I was worried about the choices he left me to make. I later learned that the best advice guides you through your challenges instead of telling you exactly what to do.

- Trust yourself. Believing in myself was not always easy during the time when I was unsure of my goals for the future. Time and experience, however, helped me determine which directions to take in my career. I used some aspects of my mentor's advice, along with what I knew about myself, to guide my decisions. Putting the two perspectives together seemed to facilitate my decision making.

- Follow up with your mentor. Always let your mentor know how you have applied his or her advice and whether the advice has proved successful. Giving feedback on your mentor's advice allows the two of you to revisit topics of discussion and delve into them more deeply. Keep in mind that both immediate and long-term follow-up are important. Nearly three years after sitting in his class, I still write to Michael and tell him about my experiences as a new teacher. I know he appreciates hearing how his advice has influenced me long after I have left his classroom. Our ongoing dialogue has also allowed the relationship to develop from a mentorship to a professional association between colleagues.

—Michael E. Matthews and Nicole I. Guarino

Helpful Hints for Mentoring

The following lists offer some helpful reminders for mentors.

Instructional Support

- Spend some time with your mentee before the start of the school year to address questions related to the school, the curriculum, administrative procedures, and so on.

- Maintain regular contact with your mentee, and ask about his or her progress. These informal conversations may offer opportunities to discuss educational issues.

- Observe the mentee's classes, then make suggestions and model instructional techniques. Such observation sessions allow you to give direct feedback related to the mentee's performance as a teacher.

- If possible, coordinate opportunities for your mentee to observe as other teachers provide instruction. In this way, the mentee learns varied teaching approaches.

- Help your mentee organize the ongoing flow of paperwork. He or she may need some time to sort and prioritize the forms and reports that need to be completed.

- When an issue arises, assist your mentee in the problem-solving process rather than solve the problem for him or her.

- Share resources, such as brief journal articles, that the mentee may find helpful for professional growth.

Affective Support

- Never underestimate the importance of trust in the mentoring relationship. Because the mentee may feel vulnerable as a new teacher, you should try to commend the positive aspects of her or his instruction and offer nonjudgmental assistance to correct problems. The mentee must believe that your job is to provide support, not to evaluate her or his performance. As a mentor, you should also always maintain confidentiality. Your mentee needs to know that she or he can discuss complex issues without repercussions.

- Be the coach without overwhelming the mentee. New teachers are often inundated with information even before they walk into their first classes. As a mentor, you should help the mentee focus on the most important issues, such as classroom management and lesson planning.

- Give personal support when your mentee feels anxious and frazzled. Assure him or her that all new teachers experience stress.

- Most of all, listen to the mentee's successes and concerns. Mentors are not expected to solve every problem or have a response for every issue. Often, the mentee benefits from simply sharing experiences with a trusted supporter.

—Joy B. Schackow, Gladis Kersaint, Tammy Rush, Virginia Harrell, Joyce McClain, and Janet Boatman

Two additional titles appear in the
Empowering the Mentor of the Mathematics Teacher series
(Gwen Zimmermann, series editor):

- ***Empowering the Mentor of the Preservice Mathematics Teacher,***
  edited by Gwen Zimmermann, Patricia Guinee, Linda M. Fulmore,
  and Elizabeth Murray

- ***Empowering the Mentor of the Experienced Mathematics Teacher,***
  edited by Gwen Zimmermann, Patricia Guinee, Linda M. Fulmore,
  and Elizabeth Murray

Please consult www.nctm.org/catalog for the availability of these titles,
as well as for a plethora of resources for teachers of mathematics at all
grade levels.

For the most up-to-date listing of NCTM
resources on topics of interest to mathematics
educators, as well as information on membership
benefits, conferences, and workshops, visit the
NCTM Web site at www.nctm.org.